AF678832

ENERGY POLICIES, POLITICS AND PRICES

WIND TURBINES

PRICE TRENDS AND EXPORT OPPORTUNITIES IN CANADA AND LATIN AMERICA

ENERGY POLICIES, POLITICS AND PRICES

Additional books in this series can be found on Nova's website under the Series tab.

Additional E-books in this series can be found on Nova's website under the E-book tab.

ENERGY POLICIES, POLITICS AND PRICES

WIND TURBINES

PRICE TRENDS AND EXPORT OPPORTUNITIES IN CANADA AND LATIN AMERICA

DERRICK A. FOSTER
EDITOR

nova publishers
New York

Copyright © 2013 by Nova Science Publishers, Inc.

All rights reserved. No part of this book may be reproduced, stored in a retrieval system or transmitted in any form or by any means: electronic, electrostatic, magnetic, tape, mechanical photocopying, recording or otherwise without the written permission of the Publisher.

For permission to use material from this book please contact us:
Telephone 631-231-7269; Fax 631-231-8175
Web Site: http://www.novapublishers.com

NOTICE TO THE READER

The Publisher has taken reasonable care in the preparation of this book, but makes no expressed or implied warranty of any kind and assumes no responsibility for any errors or omissions. No liability is assumed for incidental or consequential damages in connection with or arising out of information contained in this book. The Publisher shall not be liable for any special, consequential, or exemplary damages resulting, in whole or in part, from the readers' use of, or reliance upon, this material. Any parts of this book based on government reports are so indicated and copyright is claimed for those parts to the extent applicable to compilations of such works.

Independent verification should be sought for any data, advice or recommendations contained in this book. In addition, no responsibility is assumed by the publisher for any injury and/or damage to persons or property arising from any methods, products, instructions, ideas or otherwise contained in this publication.

This publication is designed to provide accurate and authoritative information with regard to the subject matter covered herein. It is sold with the clear understanding that the Publisher is not engaged in rendering legal or any other professional services. If legal or any other expert assistance is required, the services of a competent person should be sought. FROM A DECLARATION OF PARTICIPANTS JOINTLY ADOPTED BY A COMMITTEE OF THE AMERICAN BAR ASSOCIATION AND A COMMITTEE OF PUBLISHERS.

Additional color graphics may be available in the e-book version of this book.

Library of Congress Cataloging-in-Publication Data

ISBN: 978-1-62417-569-5

Published by Nova Science Publishers, Inc. † New York

CONTENTS

Preface

This book reviews past U.S. exports of utility-scale wind turbine nacelles and blades, and assesses the potential for future U.S. exports of these products to markets in Canada and Latin America. These markets are growing in importance and, though they are smaller than the Chinese and European markets, are significant potential export destinations for U.S. producers. The results of this analysis indicate that U.S. exports increased annually during 2007-2011, with Canada and Latin America accounting for the largest share of exports, but that the total volume of U.S. exports remains low in comparison to global exports. Markets in Ontario, Western Canada, and Mexico are growing rapidly and U.S. producers are positioned to increase exports to these places in the next few years.

Chapter 1 – This paper reviews past U.S. exports of utility-scale wind turbine nacelles and blades, and assesses the potential for future U.S. exports of these products to markets in Canada and Latin America. These markets are growing in importance and, though they are smaller than the Chinese and European markets, are significant potential export destinations for U.S. producers. The results of this analysis indicate that U.S. exports increased annually during 2007–11, with Canada and Latin America accounting for the largest share of exports, but that the total volume of U.S. exports remains low in comparison to global exports. Markets in Ontario, Western Canada, and Mexico are growing rapidly and U.S. producers are positioned to increase exports to these places in the next few years. However, U.S. producers are unlikely to benefit from installations in Quebec and U.S. exports to Brazil will be limited by trade barriers. The remaining markets in Latin America offer at least limited export opportunities, but these opportunities vary by country and depend on how the manufacturers producing in the United States decide to allocate production across their global supply chains.

Chapter 2 - Berkeley Lab has gathered price data on 81 U.S. wind turbine transactions totaling 23,850 MW announced from 1997 through early 2011.

After hitting a low of roughly $750/kW from 2000 to 2002, average wind turbine prices *doubled* through 2008, rising to an average of roughly $1,500/kW. Wind turbine prices have since declined substantially, with price quotes for transactions executed in 2010 and to date in 2011 ranging from $900-$1,400/kW depending on the manufacturer and turbine model. For example, turbines designed for lower wind speed sites – deploying higher hub heights and larger rotor diameters for a given nameplate capacity – are priced at the higher end of this range. These quotes suggest price declines of as much as 33% or more since late 2008, with an average decline closer to perhaps 20% for orders announced in 2010 (as opposed to in 2011, which has seen further price declines).

These two substantial and opposing wind turbine price trends over the past decade – and particularly the doubling in prices in the 2002-2008 period – run counter to the smooth, gradually declining technology cost trajectories that are often assumed by energy analysts modeling the diffusion of new technologies, including wind power. Understanding and explaining this notable discrepancy between theory and historical reality is the primary motivation for this work.

In: Wind Turbines
Editor: Derrick A. Foster

ISBN: 978-1-62417-569-5
© 2013 Nova Science Publishers, Inc.

Chapter 1

U.S. WIND TURBINE EXPORT OPPORTUNITIES IN CANADA AND LATIN AMERICA*

U.S. International Trade Commission

ABSTRACT

This paper reviews past U.S. exports of utility-scale wind turbine nacelles and blades, and assesses the potential for future U.S. exports of these products to markets in Canada and Latin America. These markets are growing in importance and, though they are smaller than the Chinese and European markets, are significant potential export destinations for U.S. producers. The results of this analysis indicate that U.S. exports increased annually during 2007–11, with Canada and Latin America accounting for the largest share of exports, but that the total volume of U.S. exports remains low in comparison to global exports. Markets in Ontario, Western Canada, and Mexico are growing rapidly and U.S. producers are positioned to increase exports to these places in the next few years. However, U.S. producers are unlikely to benefit from installations in Quebec and U.S. exports to Brazil will be limited by trade barriers. The remaining markets in Latin America offer at least limited export opportunities, but these opportunities vary by country and depend

* This is an edited, reformatted and augmented version of the U.S. International Trade Commission publication No. ID-032, dated July 2012.

on how the manufacturers producing in the United States decide to allocate production across their global supply chains.

INTRODUCTION

As global renewable energy markets have expanded, leaders in many countries have started looking at the renewable energy manufacturing sector as a potential export industry. In the United States, a substantial wind turbine manufacturing industry has developed in response to the growth of the domestic market, raising the question of whether this industry has significant export potential. This paper reviews past U.S. exports of utility-scale[2] wind turbine nacelles and blades and assesses the potential for future U.S. exports of these products to markets in Canada and Latin America. These markets are rapidly growing and, though they remain smaller than the large European and Chinese markets, present some of the most substantial export opportunities for U.S. producers. The results of this analysis indicate that U.S. exports of wind-powered generating sets[3] increased each year during 2007–11, but the total volume of U.S. exports remains low in comparison to global exports (box 1). U.S. producers are positioned to increase exports to Ontario, Western Canada, Mexico, and some markets in Central and South America, but are unlikely to benefit from installations in Quebec and trade barriers will limit U.S. exports to Brazil.

Box 1. Key findings

- U.S. exports of wind-powered generating sets increased each year during 2007–11, with Canada and Latin America accounting for the largest share of exports. The total volume of U.S. exports remains low in comparison to global exports of these goods.
- Markets in Ontario, Western Canada, and Mexico are growing rapidly, and U.S. producers are positioned to increase exports to these markets in the next few years.
- U.S. producers are unlikely to benefit from installations in Quebec, while U.S. exports to Brazil will be limited by trade barriers and the growth in local production.
- U.S. producers are currently supplying the Central American market, and U.S. exports to the region could grow if the pace of wind installations accelerates.

- The remaining markets in Latin America offer at least limited export opportunities. These export opportunities vary by country, and whether they are realized will depend on how the firms producing in the United States decide to allocate production across their global supply chains.

PRODUCT AND GEOGRAPHIC COVERAGE

This paper examines U.S. exports of two key components of a utility-scale wind turbine: the nacelle and blades (box 2).[4] Nacelles house the main components of the wind turbine (e.g., the gearbox and the generator). They are typically manufactured in-house by original equipment manufacturers (OEMs),[5] while blades may be either produced by the OEM or made to the OEM's specifications by an outside supplier. Project developers (the firms that develop the wind project) purchase the entire wind turbine from the OEM.[6] Therefore, exports of blades are affected by which OEMs are doing well in foreign markets and where those OEMs choose to source blades for wind turbines.

Box 2. Wind turbine nacelles and blades

Nacelles: The nacelle houses the main components of the wind turbine, such as the gearbox, generator, and shafts. According to Navigant's BTM Consult, the most commonly supplied nacelles globally in 2011 were for 1.5 to 2.5 megawatt (MW) wind turbines. Wind turbines have grown larger over time: worldwide, the average size of installed turbines rose from 1.5 MW in 2007 to 1.7 MW in 2011.

Blades: Most wind turbines have three blades. The blades are bolted to a hub attached to the front of the nacelle. Blades for 1.5 to 2.5 MW turbines typically range from 35 to 50 meters (115 to 164 feet).

Sources: David, Wind Turbines, June 2009, 2; Navigant's BTM Consult, World Market Update 2011, March 2012, 35, 52; LM Wind Web site, http://www.lmwindpower.com/Blades/Products/Blade%20summary.aspx (accessed April 13, 2012); Vestas Web site, http://www.vestas.com/en/media/brochures.aspx (accessed April 13, 2012); GE Web site, http://www.ge-energy.com/ productsands ervices/products/windturbines/ (accessed April 13, 2012).

Photo courtesy of DOE/NREL, credit Dennis Schroeder.

This paper focuses on Canada and Latin America since these markets have accounted for the largest share of U.S. exports to this point, and the growth in these markets and their proximity to U.S. producers suggest they offer further export potential.

Annual installations in Canada and Latin America are projected to rise from 2,169 megawatts (MW) in 2011 to almost 5,000 MW in 2015, an increase of more than 120 percent as compared with a projected rate of increase in global installations of about 40 percent (figure 1).[7]

While 2015 installations in Canada and Latin America will still only account for 8.3 percent of global installations—up from 5.3 percent in 2011—the growth of these markets is a significant export opportunity for U.S. producers.[8]

The large markets in China and the EU, meanwhile, offer very limited export opportunities for U.S. producers, because they are principally supplied by local producers and because many of the firms producing in the United States also have plants in one or both of these markets.[9] This paper does not assess the potential for U.S. exports to emerging markets in Asia, Africa, or other locations.

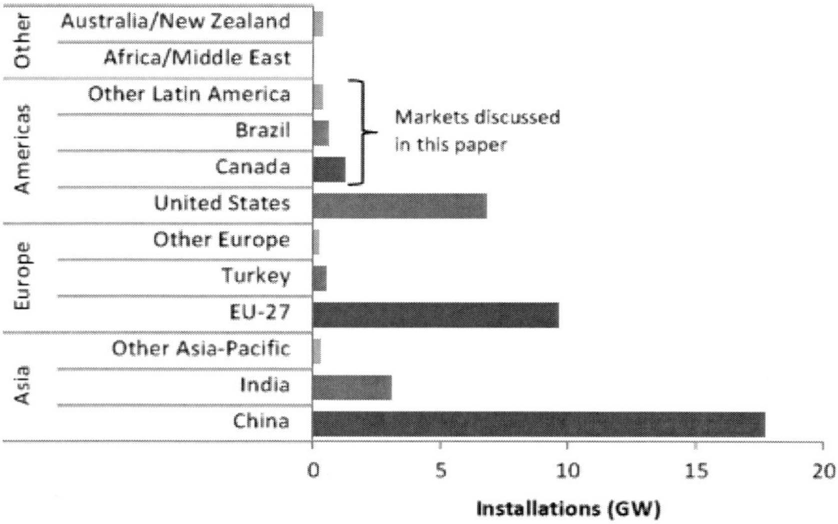

Source: GWEC, Global Wind Report: Annual Market Update 2011, March 2012, 11.

Figure 1. China, the EU, and the United States were the largest markets in 2011, but the Canadian and Latin American markets are of increasing significance.

KEY TERMINOLOGY

The production of wind turbine nacelles in the United States, as will be discussed below, is dominated by large global firms. Many of the OEMs that produce nacelles in the United States also have production in Europe, and some produce in Asia as well. In this paper, production in the United States by a firm, regardless of where the firm is headquartered, is considered U.S. production. For example, production in Colorado by Vestas, which is headquartered in Denmark, is considered U.S. production, and the Vestas units producing in the United States are considered U.S. producers.

Much of the data that are available on the wind industry, however, are firm level data. For example, data are available on all Vestas installations in Canada during 2007–11 and on total wind turbine shipments by Siemens in 2011. In order to fully understand trends and export potential for overseas markets, this paper will look at firm-level data before examining the export prospects for U.S. industry. As will be discussed, many firms have a choice of whether to supply a foreign market from their U.S. plant or from one of their plants in another country, so a wind turbine contract between a firm with U.S. production and a project developer in Canada or Latin America does not always lead to a U.S. export. This paper will discuss the size of the "project pipeline," sometimes also referred to as just the "pipeline," for firms with U.S. production. For the purposes of this paper, a firm's project pipeline is orders and planned installations[10] of the company's wind turbines and is measured in MW. This information is based on a number of sources such as company press releases, media reports, and project documents and Web sites.

Box 3. Key terminology as used in this report

- *Feed-in tariff (FIT):* A policy mechanism that requires the purchase of electricity from a renewable energy system at a specified price for a certain period of time. The FIT rates are often lowered over time as the cost of renewable energy systems decline.
- *Power purchase agreement (PPA):* A long-term agreement between a utility and an independent power producer to purchase electricity.
- *Project pipeline:* A firm's project pipeline is orders and planned installations of the company's wind turbines. It is measured in MW.
- *Renewable portfolio standard (RPS):* A mandate that a certain percentage of renewable energy come from renewable sources by a particular date.
- *Tender:* In some markets, utilities select wind projects through a tender process. A utility indicates that they will contract for a certain amount of electric generating capacity, project developers submit bids, and the utility selects the winning bid based on low price and/or other considerations.
- *U.S. producer:* A company producing in the United States, regardless of where the parent firm is headquartered.
- *U.S.-based firm:* A company headquartered in the United States.
- *Utility-scale:* In this paper, utility-scale wind turbines are those rated at 100 kilowatts (kW) or more.
- *Wind-powered generating set:* Refers to the complete nacelle and any items imported with the nacelle. Used for trade data. Wind-powered generating sets are included in 8502.31 in the international Harmonized Commodity Description and Coding System (HS).
- *Wind turbine:* The complete turbine, including the nacelle, hub, blades, and tower.

In this paper, the size of a country's wind market is defined based on MW installed. Annual installations are the MW installed in a single year, while cumulative installations are the sum of all annual installations. Wind turbine installations are not always correlated with wind turbine shipments, since wind turbines are often shipped in one year but not installed until the next year.

Several terms are used in this paper in reference to wind turbines (box 3). In this paper, the term wind turbine refers to the entire structure, including the complete nacelle, hub, blades, and tower. When the terms nacelle and blade are used, they refer to the specific components referenced and not to the whole turbine. Finally, the trade term wind-powered generating sets will be used to refer to imports and exports. This term has a specific meaning for trade data, which is the complete nacelle and any items imported with the nacelle. If blades and hubs, for example, are imported in the same shipment as nacelles, they are considered part of the wind-powered generating set. If they are imported separately from the nacelle, they are not included in the trade data for wind-powered generating sets.

U.S. Market Uncertainty

This paper's assessment of the export opportunities for the U.S. wind industry is based on the current production capacity in the United States. The U.S. industry, however, is facing a potentially large drop in domestic demand due to the pending expiration of several federal tax credits/incentives for wind, including the production tax credit (PTC), bonus depreciation, and investment tax credit (ITC) and grant in lieu of the ITC (box 4). State renewable portfolio standards (RPSs) remain a driver of wind installations, but without federal government policies are not projected to sustain installations at the current levels.

The U.S. wind turbine market averaged more than 7 gigawatts (GW) in installations during 2007–11 and is projected to be between 8 and 12 GW in 2012 as companies rush to complete projects before incentives expire. However, installations are forecast to fall to between 0.5 GW and just over 2 GW in 2013 and to be less than 3 GW in 2014 if the PTC is not renewed (figure 2).[11]

U.S. nacelle capacity is more than 10 GW, so a market of this size may not be large enough to sustain some of the U.S. manufacturing plants.[12]

Box 4. Select U.S. government policies for wind

- Bonus depreciation: Wind projects were eligible for 50 percent bonus depreciation in their first year during 2008–
- September 2010, 100 percent bonus depreciation during September 2010–2011, and 50 percent in 2012.
- Investment tax credit (ITC)/ITC grant: Wind projects were also made eligible for the 30 percent ITC in 2009 or could elect to receive a cash grant in place of the ITC. In order to be eligible for the ITC, wind projects must be completed by the end of 2012. To be eligible for the cash grant, projects must have started construction by the end of 2011 and must be completed by the end of 2012.
- Production tax credit (PTC): A tax credit based on the amount of electricity generated by a wind project over its first ten years. The tax credit is adjusted for inflation and in 2010 was 2.2 cents per kilowatt-hour (kWh). Currently, projects must be completed by the end of 2012 to receive the tax credit.

Source: USDOE, Loan Programs Office Web site, https://lpo.energy.gov/?pageid=41; Wiser and Bolinger, 2010 Wind Technologies Market Report, June 2011, 61–62; Platzer, U.S. Wind Turbine Manufacturing, September 23, 2011, 25–30.

Note: Wind project developers must choose between the PTC, ITC, and ITC grant. Wind projects also became eligible for loan guarantees under the American Recovery and Reinvestment Act of 2009. Four wind projects received full or partial guarantees.

Uncertainty over the PTC may lead to more exports in the short term since companies are looking at export markets to help offset the decline in U.S. demand, but an extended period without the PTC could lead to fewer long-term exports.[13]

The export opportunities discussed here are unlikely to offset all of the U.S. sales decline and a lack of demand could lead to a contraction in U.S. production capacity.

This paper does not analyze the impact of U.S. market scenarios on production capacity, but it is possible that expiration of the PTC could lead to a smaller industry that exports fewer products.

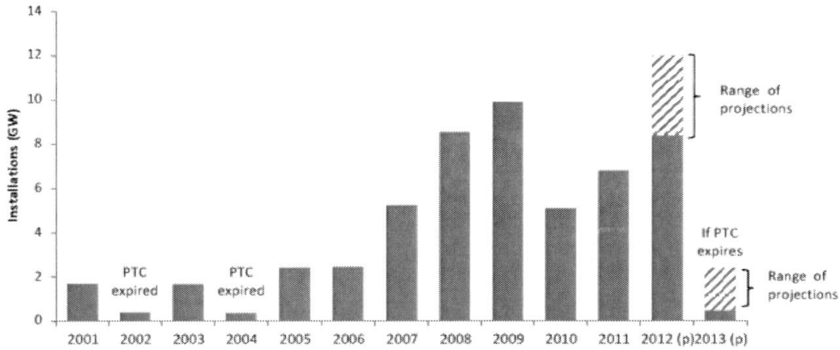

Sources: Navigant Consulting, Impact of the Production Tax Credit, December 12, 2011, 13; Zindler, written statement to the Senate Committee on Finance, March 27, 2012, 2; Kaplan, "United States Wind Market," November 2, 2011, 6; Voorhis, "Energy Industry Fears," April 5, 2012; AWEA, U.S. Wind Industry Fourth Quarter 2011 Market Report, January 2012, 1; AWEA, U.S. Wind Industry Year-End 2010 Market Report, January 2011, 1; AWEA, Annual Wind Industry Report: Year Ending 2008, 2009, 4; AWEA, Year-End 2009 Market Report, January 2010, 2.

Note: p: projected installations.

Figure 2. U.S. wind installations are projected to fall in 2013 if the production tax credit is not renewed.

U.S. Wind Industry Expanded in Response to Domestic Market Growth

There has been substantial investment in U.S. nacelle production by foreign OEMs, particularly those based in the EU, in response to growing U.S. demand (table 1).[14] In addition, several U.S.-based companies also produce domestically. In 2011 all three firms with at least 1,000 MW in installations in the United States, and six of the top nine firms in the U.S. market (considering REpower and Suzlon separately), assembled nacelles in the United States.[15] In addition, a number of companies with smaller markets shares are producing nacelles in the United States. This growth in production was enough to make the United States the third largest nacelle producer in 2011, but U.S. production was still significantly lower than that in China and Europe (box 5).

Three OEMs (Gamesa, Siemens, and Vestas) have established blade production in the United States. In addition, several independent suppliers of blades to OEMs produce in the United States. TPI Composites and Molded Fiberglass supply GE from their U.S. plants. LM Wind Power likely supplies multiple OEMs from its U.S. plants. It is the largest global independent blade supplier and had a 27 percent market share in North America in 2010.[16]

Table 1. U.S. nacelle and blade producers, March 2012

Company	Headquarters	2011 U.S. MW installed	U.S. production	Production capacity		Exported in past?
				MW	Units	
OEM						
Acciona Windpower	Spain	0	Nacelles: Iowa	675	450	Yes
Aeronautica	United States	0.8	Nacelles: New Hampshire	N/A	N/A	N/A, but plan to export from U.S. plant
Alstom	France	39.6	Nacelles: Texas	800	N/A	No, but plan to serve the North American market from U.S. plant
Daewoo Shipbuilding & Marine Engineering Company (DSME)/ DeWind[a]	Korea	0	Nacelles: Texas	600	300	Yes
Gamesa	Spain	152	Nacelles: Pennsylvania Blades: Pennsylvania	1,200 N/A	N/A N/A	Yes
General Electric Co. (GE)	United States	2,006.1	Nacelles: California, Florida, and South Carolina	5,024	3,140	Yes
Nordex	Germany	287.5	Nacelles: Arkansas	750	300	N/A, but plan to serve Canada and Latin

Company	Headquarters	2011 U.S. MW	U.S. production	Production capacity		Exported in past?
		installed		MW	Units	
						America from U.S. plant
Northern Power Systems	United States	2.3	Nacelles: Michigan, Vermont Blades: California	N/A N/A	N/A N/A	Yes
Siemens	Germany	1,233.1	Nacelles: Kansas Blades: Iowa	1,500 1,610	650 2,100	N/A, but plan to export from U.S. plants
United Technologies Corp. (UTC)/Clipper Windpower[b]	United States	257.5	Nacelles: Iowa	>1,000	>400	Yes
Vestas Wind Systems	Denmark	1,969.4	Nacelles: Colorado Blades: Colorado (2 plants)	N/A 3,000	1,500 3,800	Yes
Independent blade suppliers						
Energetx Composites	United States	N/A	Blades: Michigan	N/A	N/A	No blade exports, did export blade molds
LM Wind Power	Denmark	N/A	Blades: Arkansas, North Dakota	N/A	N/A	Yes
Molded Fiberglass	United States	N/A	Blades: South Dakota, Texas	N/A	N/A	Yes
TPI Composites	United States	N/A	Blades: Iowa	N/A	N/A	N/A

Sources: Compiled from press releases, media reports, company documents and Web sites, interviews with industry representatives, and AWEA publications. GE production capacity is from the Bloomberg New Energy Finance database.

Notes: N/A: not available. Does not include planned plants. For Siemens, the figure for blade production capacity is based on its expected annual number of blades produced as of September 2011, and the figure for MW is based on the assumption that all blades are used for its 2.3 MW wind turbine. GE's production capacity is based on the average-size GE wind turbine installed in the United States in 2011, which was 1.6 MW. Nacelles for Northern Power Systems' 2.5 MW turbines are manufactured by Merrill Technologies Group in Michigan, and the nacelles for the DeWind wind turbine are produced by TECO Westinghouse in Texas. In some cases production capacity is announced capacity, and companies may not have reached the announced production levels. The DeWind production rate is TECO Westinghouse's production rate for one shift per year. Construction of a planned Mitsubishi nacelle plant was suspended in 2012. Northern Power Systems acquired Knight & Carver's blade group. Nordic Windpower announced plans in 2010 to move its manufacturing plant from Pocatello, Idaho to Kansas City, Missouri. Information is not available on Nordic's current status, so Nordic is not included in this list. Several other firms, including Blade Dynamics and Global Blade Technology, have set up blade plants, but information is not available on whether they have started commercial production.

[a] Subsequently referred to as "DeWind."

[b] Subsequently referred to as "Clipper" or "Clipper Windpower."

Box 5. Global wind industry in 2011

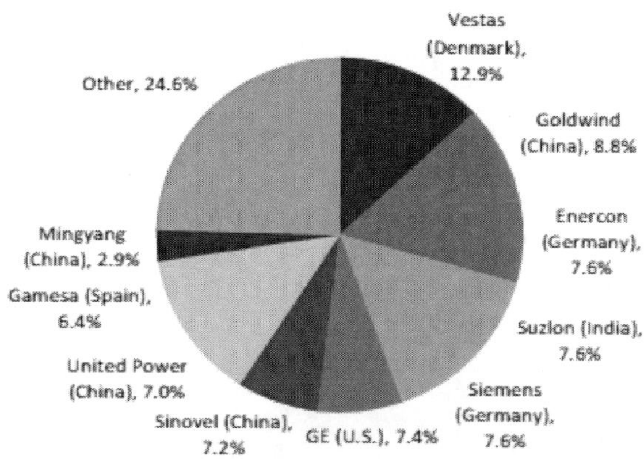

Source: North American Windpower,"Report Reveals Top 10 Wind Turbine Manu facturers for 2011," March 26, 2012.
Notes: MAKE Consulting estimate of 2011 installations. Total not available. Headquarters location in parentheses.

EU- and China-based firms were the leaders in 2011 global installations.

Most of the leading wind turbine suppliers in 2011 were based in the EU and China.

EU-based firms have significant production in their home market from which they supply both domestic and foreign markets. These firms have also localized production in key markets such as Brazil, China, and the United States.

China-based firms produce only in China and have primarily supplied the domestic market, though many are actively seeking to expand their foreign sales. U.S.-based GE and India-based Suzlon are also among the leading global producers.

China was the leading producer of nacelles in 2011, followed by the EU, the United States, and India. The EU, however, was the dominant global exporter of wind-powered generating sets in 2011 due to its large domestic industry and sales by EU firms in many overseas markets. China was second, followed by the United States. India and Japan have also been significant exporters in past years, though their exports were low in 2011 because of a drop in shipments to the United States.

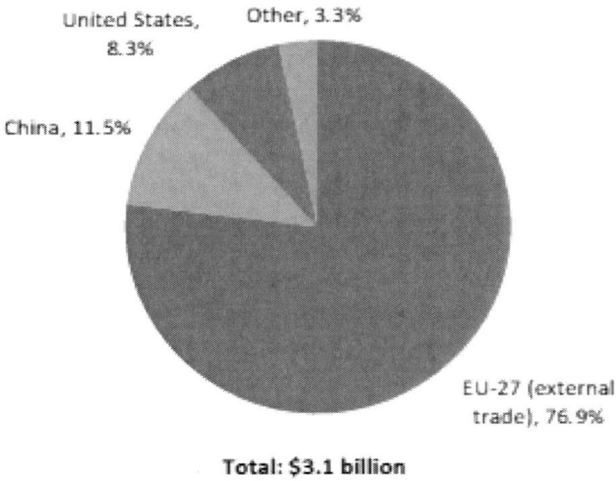

Source: GTIS,Global Trade Atlas database (accessed July 23, 2012).

Note: U.S. export data from GTIS may differ slightly from U.S. export data discussed elsewhere in this paper.

The EU was the dominant global exporter of wind-powered generating sets in 2011.

Source: Information on the leading nacelle producing countries is based on estimates by USITC staff.

U.S. EXPORTS SIGNIFICANTLY INCREASED DURING 2007–11

U.S. exports of wind-powered generating sets[17] rose substantially during 2007–11, with Canada and Latin America accounting for the largest share of exports (figure 3 and table 2).[18] The increase in U.S. exports is driven principally by three factors: the increase in the number of firms producing in the United States and the related growth in U.S. production capacity; growing markets in Canada and Latin America; and the competitiveness of U.S. firms in nearby markets. A majority of U.S. nacelle producers have exported in the past and those that have not exported have indicated that they intend to supply foreign market from their U.S. plants (table 1). Despite the increase in U.S. exports, they accounted for only 8 percent of global exports of wind-powered generating sets in 2011.[19]

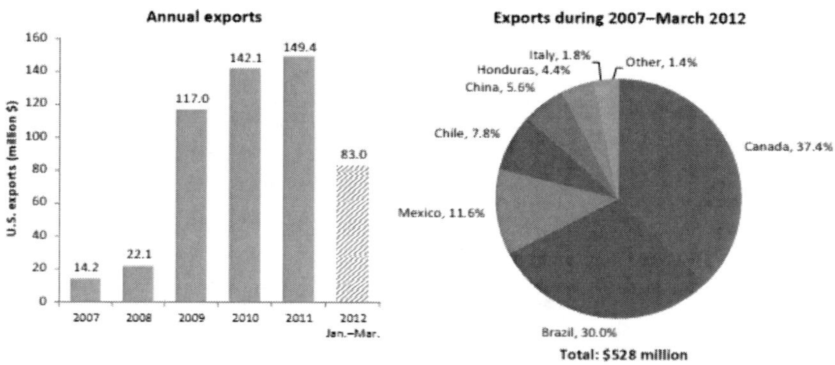

Source: USITC DataWeb/USDOC (accessed April 27, 2012).

Figure 3. U.S. exports of wind-powered generating sets rose annually during 2007–11 and were primarily to Canada and Latin America.

One of the main reasons that U.S. producers are competitive in nearby markets is that transportation costs are lower than for producers located farther away.[20] U.S. data illustrate the expense of exporting from Europe to the Americas. In 2011, for instance, insurance, freight and other charges for wind-powered generating sets exported from Denmark to the Houston-Galveston customs district averaged $39,000 per wind-powered generating set (2.6 percent of the average import value); to the Minneapolis customs district, costs were $39,000 (3.0 percent); and to the Columbia-Snake customs district, $63,000 (5.7 percent).[21]

Table 2. U.S. exports of wind-powered generating sets, 2007–March 2012, million $

	2007	*2008*	*2009*	*2010*	*2011*	*Jan.–Mar. 2012*	*Total*
Americas Canada	0.0	0.0	77.7	107.9	11.8	0.0	197.4
Brazil	0.0	5.9	0.0	0.0	99.6	52.7	158.2
Mexico	0.0	0.0	18.1	24.4	11.8	6.7	61.0
Chile	0.0	0.0	18.1	0.0	0.0	22.9	41.1
Honduras	0.0	0.0	0.0	0.0	23.4	0.0	23.4
Argentina	0.0	0.0	1.2	0.0	0.0	0.0	1.2
Saint Kitts and Nevis	0.7	0.0	0.0	0.0	0.0	0.0	0.7
Bahamas	0.0	0.0	0.0	0.0	0.6	0.0	0.6
Other China	12.8	16.2	0.0	0.0	0.6	0.0	29.6
Italy	0.0	0.0	1.2	7.6	0.8	0.0	9.6
Korea	0.0	0.0	0.0	2.1	0.8	0.0	3.0
Spain	0.0	0.0	0.7	0.0	0.0	0.0	0.7
Japan	0.6	0.0	0.0	0.0	0.0	0.0	0.6
United Kingdom	0.0	0.0	0.0	0.0	0.0	0.6	0.6
Total	14.2	22.1	117.0	142.1	149.4	83.0	527.7

Sources: USITC DataWeb/USDOC (accessed April 27, 2012).

Notes: Exports to Honduras in 2011 were incorrectly categorized as foreign exports (re-exports or goods that are imported into the U.S. and then exported in essentially the same condition), so they are included here with domestic exports. U.S. exports totaled $0.8 million during Jan.–Mar. 2011.

Industry representatives indicate that the ability to access competitive financing from the Export-Import Bank of the United States (Ex-Im Bank or Ex-Im) can benefit U.S. firms that export, especially in Latin America where financing projects can be challenging. Ex-Im Bank has a target, set by Congress,[22] to provide 10 percent of its export financing to renewable energy or energy efficient products. It has established an Environmental Exports Program under which it offers enhanced financing (e.g., extended repayment terms) and other incentives (e.g., the ability to lock in interest rates when the loan is approved) for renewable energy. In addition, for renewable energy Ex-Im Bank can provide financing for projects in Canada. At least four firms have accessed Ex-Im financing for projects in Latin America, and other firms are reportedly considering doing the same (table 3).

Table 3. Announced Ex-Im Bank funding for wind exports, 2009–June 2012

Company	Announced	Project	Number of	Turbine rating	Total	Ex-Im financing
		location	turbines	(MW)	(MW)	(million $)
Gamesa (nacelles) and	August 20, 2010	Hondura s	51	2	102	159
LM Wind Power (blades)						
LM Wind Power (blades)	May 31, 2012	Brazil	N/A	N/A	N/A	32
Northern Power Systems	May 26, 2011	Italy	55	0.1	5.5	22
UTC/Clipper Windpower	November 24, 2009	Mexico	27	2.5	67.5	81

Sources: Ex-Im Bank, "American Windpower Firm Grows," November 24, 2009; Ex-Im Bank, "Export-Import Bank Provides $159 Million," August 20, 2010; Ex-Im Bank, "Vermont Manufacturer," May 26, 2011; Ex-IM Bank, Summary of Minutes, May 31, 2012.

Note: N/A: Not applicable.

Local-content requirements are likely the most significant barrier to U.S. exports of nacelles and blades. Several of the largest markets, including Ontario, Quebec, and Brazil, have implemented requirements to source locally, though the impact of these requirements will likely vary depending on the design of the rules.[24] A few countries in the Americas, including Brazil, maintain high tariffs on wind-powered generating sets that may affect U.S. exports, but tariffs in the region generally remain low. U.S. producers may also have a competitive advantage in countries where U.S. products enter duty free due to free trade agreements.[25]

CANADA COULD BECOME A SIGNIFICANT U.S. EXPORT MARKET

Overview

Policies at the provincial level in Canada are driving a rapid expansion of the Canadian wind turbine market and creating export opportunities for U.S.

producers of nacelles and, to a lesser extent, blades. Firms with U.S. nacelle production, like Acciona, GE, Siemens, and Vestas, have large project pipelines in Canada. Many of these firms have supplied Canadian projects from the United States in the past and intend to supply future projects from the United States. There is also the potential for additional blade exports to Canada, with at least one firm increasingly supplying the Canadian market from its U.S. plant. U.S. exports to Canada may be limited, however, by local content requirements in Ontario and Quebec.

Market

Canada is the largest wind market in the Americas after the United States, with 1.3 GW in installations in 2011 (figure 4).[26] The market is expected to grow further in the next few years, with annual installations projected to reach 2 GW in 2013 and 2.5 GW in 2014.[27] Three provinces—Ontario, Quebec, and Alberta—accounted for 72 percent of cumulative Canadian wind installations through the end of 2011, with Ontario alone accounting for 37 percent.[28]

These provinces also account for the largest share of new project development activity in Canada, though installations are also rising in other provinces.[29] In British Columbia, wind developers are hopeful that wind-generated electricity will be in demand as a result of the provincial government's need for new electricity generation to supply provincial efforts in developing liquid natural gas plants, shale gas resources, and mining activities.[30]

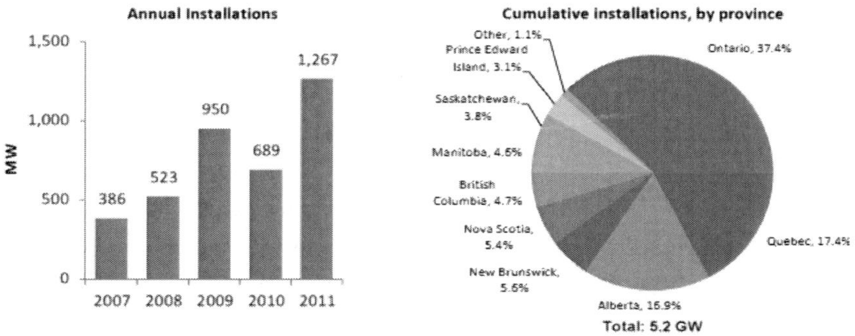

Source: CANWEA, Powering Canada's Future, December 2011.

Figure 4. Wind installations in Canada rose significantly during 2007–11.

The province's power company, BC Hydro, however, has awarded contracts for only 0.5 GW of new wind capacity.[31]Canada's national energy policy is market oriented, and thus wind installations in Canada are driven by renewable energy goals and policies at the provincial level (table 4).[32] To accomplish their goals of increasing power from renewables, the provinces, with the exceptions of Alberta and Manitoba, have established RPSs, offered or tendered contracts for electricity generated by wind power or other renewables through provincial-owned electric companies, and introduced feed-in tariffs (FITs).[33] RPSs are often used in combination with other policies, such as tendering and FITs. Ontario has the most comprehensive FIT in North America for renewable energy.[34] In September 2009, Ontario launched its FIT, which is included in the province's Green Energy and Green Economy Act of 2009. Among the goals of the FIT is to help Ontario phase out coal-fired electricity generation by 2014 and create jobs in renewable energy industries. By the end of September 2011, more than 3 GW of wind projects were being developed under the Ontario Power Authority's FIT program.[35]

The provinces of Quebec and Ontario have included domestic-content requirements in their wind programs in order to encourage economic development; however, wind turbine producers contend that such requirements act as barrier to entry and raise costs.[36] The Province of Quebec, through the province's utility company, Hydro-Québec, required 40 percent to 60 percent regional content for wind projects in its first wind tender in 2003. In the 2005 second wind tender and the 2009 third wind tender, Hydro-Québec required that a minimum of 30 percent of the content of the wind turbines must be produced in manufacturing plants in the regional county municipality of Matane and the administrative region of Gaspésie-Îles-de-la-Madeleine in Quebec and a minimum of 60 percent of the costs of the wind farms must come from Quebec.[37]

Under Ontario's FIT program, wind projects over 10 kilowatts (kW) that started commercial operation during 2009–11 must have a minimum provincial domestic content level of 25 percent. The content requirement rose to 50 percent for projects starting in 2012 and later.[38] Minimum domestic content percentages are set out for 18 manufacturing inputs or activities areas carried out in Ontario. The largest domestic content percentages are for the wind turbine blades, gearbox or generator (depending on whether the turbine is direct drive), steel for towers, grid connection transformers, and construction costs and on-site labor.[39] In September 2010, Japan requested consultations with Canada under the World Trade Organization (WTO) dispute settlement process regarding domestic content requirements in Canada.

Table 4. Canada's installed wind energy capacity (MW), December 2011, and principal wind energy goals by provincial jurisdiction

Province	Cumulative capacity (MW)	Provincial wind energy goals	Major policy used
Alberta	891	No provincial target for wind.	Planned transmission improvements.
British Columbia	248	No provincial target for wind. During 2007-2012, set a goal of electricity selfsufficiency by 2016 with 93 percent from renewable energy sources. In February 2012, set a goal to develop electricity sources to power natural gas and mining projects. These sources may include wind power.	Tendering by BC Hydro for large projects; Standing Offer Program for projects up to 10 MW. Feed-in tariff proposed in 2010, but not yet adopted.
Manitoba	242	1,000 MW by 2016.	Tendering by Manitoba Hydro.
New Brunswick	294	In 2006, set an RPS of 10 percent of provincial utility electricity (kW) sold from renewables by 2016, with approximately 400 MW from wind. In October 2011, RPS raised to 40 percent of NB Power total sales by 2020, but this includes imported wind power, and improved wind power infrastructure planned.	Tendering by NB Power. Feed-in tariff for community-based wind projects with a capacity less than 15MW established in February 2010.
Newfoundland	55	80 MW (maximum electricity from wind that the grid can accept) on the island of Newfoundland.	Tendering by Newfoundland and Labrador Hydro, a subsidiary of Nalcor Energy.
Nova Scotia	286	25 percent renewable energy by 2015 (from 1,300 gigawatt-hour (GWh)/year to 3,000 GWh/year); 40 percent	RPS in Renewable Electricity Regulations of 2010; feed-in tariff system for communitybased wind

Table 4. (Continued)

Province	Cumulative capacity (MW)	Provincial wind energy goals	Major policy used
		renewable energy by 2020.	projects; tendering for half of all large- and medium-size renewable energy projects for independent power producers.
Ontario	1,970	10,700 MW of renewable energy by 2018.	Feed-in tariff.
Prince Edward Island (PEI)	164	RPS in Renewable Energy Act of 2005 for 15 percent of total generation by 2010. PEI Energy Accord, effective March 2011, sets a goal of 33 percent of electricity consumption from wind generation by 2016.	RPS in Renewable Energy Act of 2005 and RPS in PEI Energy Accord.
Quebec	918	4,000 MW by 2015; additional 100 MW of wind power for every 1,000 MW of new hydro power.	Tendering by Hydro-Québec.
Saskatchewan	198	Add 175 MW during 2010–15	Tendering by SaskPower.
Yukon	1	Increase Yukon renewable energy use by 20 percent during 2008–2020 from 297 GWh/year to 356 GWh/year. By 2012, create RPS for Yukon Energy Corp. for 90 percent clean or renewable energy.	RPS for Yukon Energy Corp. in 2012.

Sources: CANWEA, Powering Canada's Future, March 2012; various provincial Web sites.

Note: GWh: gigawatt-hour.

The Japanese complaint alleges that Ontario's FIT discriminated against imported renewable energy products relative to those made in Ontario.[40] In

August 2011, the EU requested consultations within the WTO on the same issue. Both Japan and the EU requested the formation of dispute panels.[41]

Currently, Canada does not have a national, funded renewable energy incentive program. The major Canadian national program was the ecoENERGY program, enacted in 2007 and terminated in 2010.[42] The program paid renewable energy providers 1 cent (before tax) per kilowatt-hour (kWh) for the first 10 years of commercial renewable power production. Projects had to be completed between April 1, 2007, and March 31, 2011, so payments made by the program under these contracts will continue through March 31, 2021. Two federal tax provisions also aid the development of wind in Canada. The first is a tax program that accelerates the capital cost allowance (CCA) on clean energy generation equipment and can be applied to assets acquired from February 23, 2005 to December 31, 2019.[43] The second is the Canadian Renewable and Conservation Expense (CRCE), introduced in March 1996.[44] The CRCE was established to place the tax treatment of renewable energy on a parity with the treatment given to the petroleum, natural gas, and mining industries.[45]

Market Competition

GE, Enercon, Siemens, and Vestas were the leading suppliers to the Canadian market in 2007–11, accounting for a combined 98 percent of wind turbine installations (figure 5). Acciona has developed a number of wind projects in Canada using turbines produced by other companies, but completed its first wind project using its own turbines in 2011. Other companies with installations in Canada during this period include Leitwind, Northern Power Systems, REpower, and Vensys.[46]

While four OEMs dominated the market during 2007–11, competition is increasing. Examining wind projects under construction or in development in four provinces—Alberta, British Columbia, Ontario, and Quebec—indicates that as of March 2012 seven firms each had project pipelines of at least 100 MW in the four provinces combined (figure 6).

Competition among OEMs, however, varies significantly by province, depending on the characteristics of the market and factors such as local-content requirements. In Quebec, competition is fairly limited. Quebec's second wind tender or request for bids (in 2005, for 2,000 MW, selections announced in 2008) resulted in bids including turbines from five manufacturers.

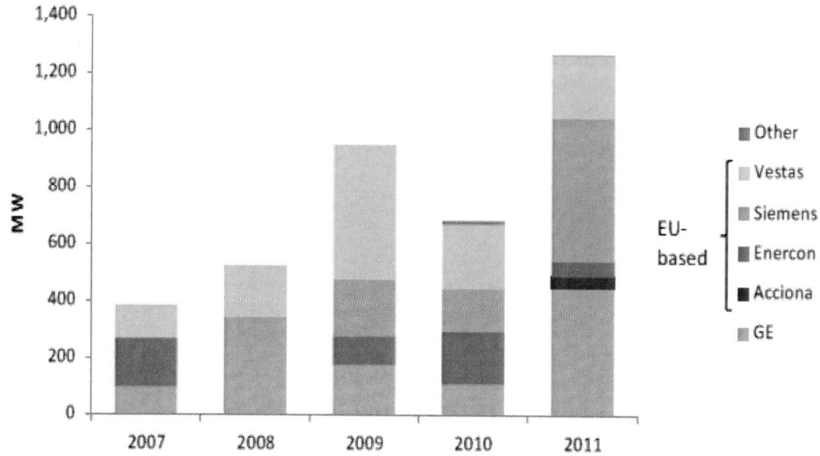

Source: CANWEA Web site, http://www.canwea.ca/farms (accessed April 2, 2011).

Figure 5. The Canadian market has been dominated by GE and EU-based producers.

Its third wind tender (in 2009, for 500 MW, selections announced in 2010) resulted in bids from project developers that included four manufacturers. At least one firm chose not participate in the Quebec tenders because the market was not large enough to warrant meeting local-content requirements.[47]

Under the Ontario FIT, which also includes a local-content requirement, the leading suppliers are GE, Siemens, and Vestas, while leading Quebec suppliers Enercon and REpower have the fourth- and fifth-largest project pipelines.

Alberta, meanwhile, has a completely deregulated utility market and projects in development there are being supplied by eight OEMs, of which Vestas and Acciona have the largest pipelines.

British Columbia's call for proposals, meanwhile, resulted in the selection of projects that are likely to use wind turbines from either Vestas or GE. Chinese wind turbine manufacturers have not made major inroads in the Canadian market yet, despite active efforts by firms such as Goldwind and Sinovel to increase their sales.[48]

Similarly, with the exception of sales by REpower, the Germany-based subsidiary of Suzlon, Indian producers have had few sales in Canada.[49]

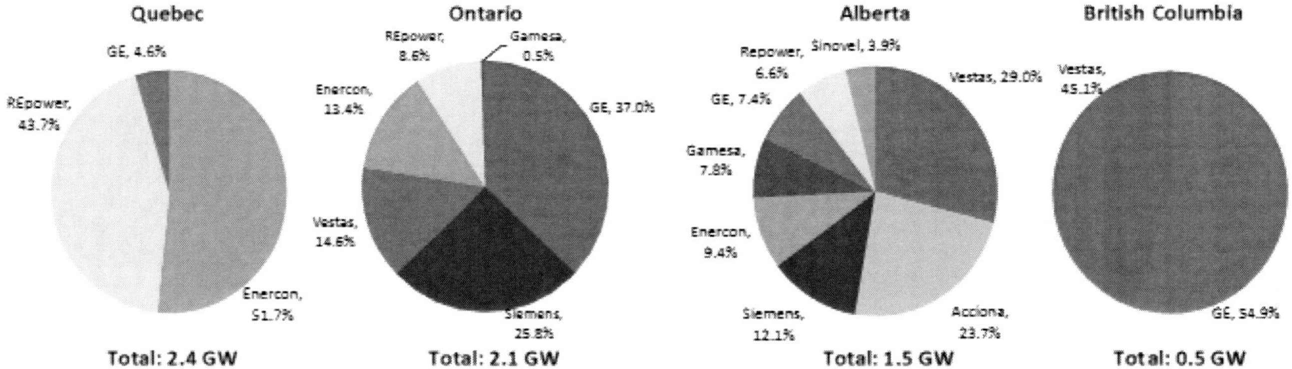

*EU-based firms are in shades of red, U.S.-based firms in blue, **and** Asia-based firms in purple.*

Sources: The list of projects is from the Ontario Power Authority's list of FIT contracts offered, Alberta Electric System Operator's active project list, and BC Hydro and Hydro-Québec information on awarded contracts under wind tenders. For Quebec, information on equipment suppliers is from the Hydro-Québec Web site; for all others it is based on media reports, press releases, company Web sites, and project documents.

Notes: The charts above only include projects for which USITC staff could identify the OEM that the project developer has contracted with or intends to use to supply the project. Therefore, the actual pipeline for at least some firms could be even larger. In addition, some project documents may list a particular OEM as the supplier, but the project developer may not have a contract with this firm and may ultimately choose to use a different supplier. Therefore, the project pipeline of firms shown here could change slightly.

U.S. EXPORT OPPORTUNITIES

The United States exported about $200 million in wind-powered generating sets to Canada during 2007– 11. In total, 164 nacelles were exported, all during the last three years of this time period. The U.S. producers that exported to Canada include Acciona, GE, and Northern Power Systems.[50] Canada's imports from the United States, however, were significantly lower than those from the EU. Canadian imports of wind-powered generating sets from the EU totaled $2.1 billion during 2007–11 and accounted for 83 percent of imports by value (figure 7). The largest source of imports was Denmark, which accounted for $1.6 billion, followed by Germany at $540.4 million.[51]

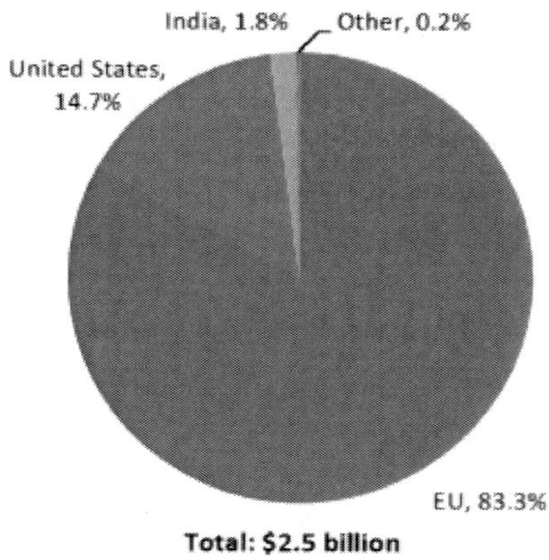

Source: GTIS, Global Trade Atlas database (accessed April 2, 2012).
Note: Based on Canadian import data, which may differ from U.S. export data discussed above.

Figure 7. The EU and the United States were the leading suppliers to Canada during 2007–11.

The large share accounted for by EU exports reflects the location of the leading suppliers' manufacturing plants. Three of the top four suppliers to the Canadian market—Enercon, Siemens, and Vestas—are EUbased firms that did not have U.S. nacelle production for most of this time period.

Siemens and Vestas did not open their U.S. nacelle plants until the second half of 2010 and Enercon does not have U.S. nacelle production.[52] GE was the second largest supplier to Canada during this time period, but established nacelle production in Quebec due to local content requirements.[53]

Based on an analysis of trade data, it appears that GE supplied nacelles for projects in Quebec from this plant, but supplied most nacelles for projects outside of Quebec from its U.S. plants.[54]

The U.S. industry is positioned to supply more nacelles to Canada in the future due to the establishment of U.S. plants by Siemens and Vestas, more firms with U.S. production entering the market, the fact that most of GE's future pipeline is outside of Quebec, and the growth of the Canadian market. The project pipeline for firms with U.S. production totals about 3.6 GW. This includes three firms with projectpipelines of more than 700 MW—GE, Siemens, and Vestas—and two firms with pipelines of more than 100 MW—Acciona and Gamesa. Most export opportunities for U.S. produces are in Ontario, Alberta, and British Columbia (figure 8).

Though Ontario has a local content requirement, this can be met without producing nacelles in Ontario, as long as blades or at least some nacelle components are sourced locally (box 6).

Quebec has a large market, but there are no current export opportunities to this market for U.S. nacelle producers since GE supplies that market from its plant in Canada and the winning suppliers in the last two wind tenders—Enercon and REpower—do not have U.S. nacelle plants.

The firms with project pipelines in Canada are likely to source at least a portion of the nacelles for the Canadian market from the United States (table 5). GE and Acciona both exported nacelles from the United States to Canada during 2007–11 and could again export nacelles for some of the projects in their pipeline.[55]

GE has tended to supply a larger share of demand outside of Quebec from the United States, and will likely continue to supply some Canadian projects from the United States.

DeWind and Gamesa both indicated that they will export turbines for 10 MW projects.[56] Vestas intends to serve the Canadian market from its nacelle plant in Colorado and nacelles for at least three upcoming projects totaling 345 MW will be produced in Colorado.[57]

Siemens also plans to meet a substantial portion of Canadian demand from its U.S. nacelle plant in Hutchinson, Kansas. According to one Siemens representative, "Once Hutchinson is up and running, we expect to supply the majority of all projects we do in the U.S. and Canada from that facility."[58]

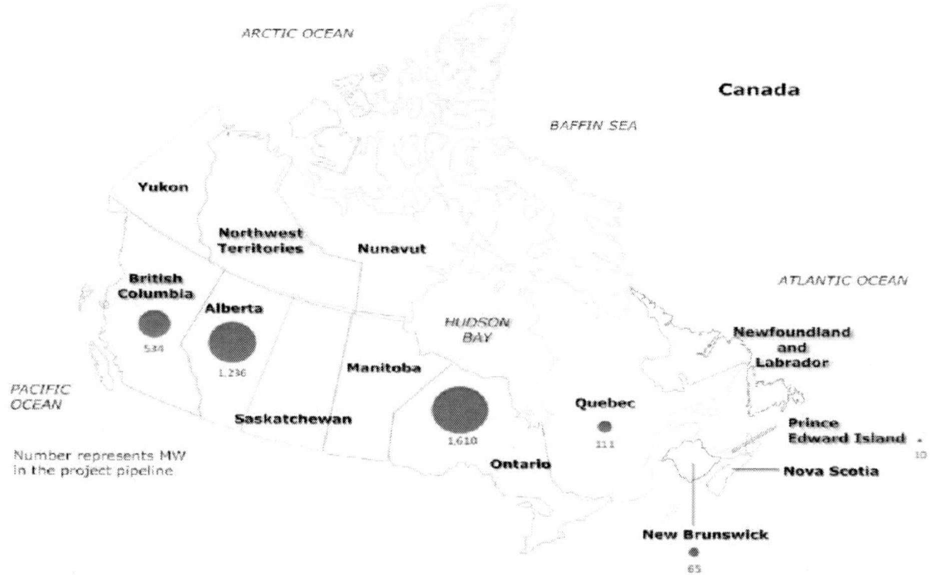

Source: The list of projects is from the Ontario Power Authority's list of FIT contracts offered, Alberta Electric System Operator's active project list, and BC Hydro and Hydro-Québec information on projects awarded contracts under wind tenders. For all other provinces, it includes projects under construction or in development identified by USITC staff. For Hydro-Québec information on equipment suppliers comes from the utility's Web site; for all other provinces it is based on media reports, press releases, company Web sites, and project documents.

Notes: This figure represents projects under construction or in development that are planning to use turbines from OEMs that have nacelle plants in the United States. It only includes projects for which USITC staff could identify the OEM that the project developer has contracted with or intends to use to supply the project. Therefore, the actual pipeline for firms could be even larger. In addition, some project documents may list a particular OEM as the supplier, but the project developer may not have a contract with this firm and may ultimately choose to use a different supplier. Therefore, the project pipeline of firms shown here could change slightly.

Figure 8. For firms with U.S. nacelle plants, the largest project pipelines are in Ontario, Alberta, and British Columbia, April 2012.

Box 6. Local-content rules in Ontario

In Ontario, the local-content rules establish "qualifying percentages" for specific components and services that are part of the wind project. Local content must be 50 percent as calculated using these pre-established percentages. As a result, there are a number of different ways that suppliers can reach the mandatory 50 percent local-content percentage. Sourcing everything but the blades, the nacelle, and nacelle components from Ontario results in a local content share of more than 40 percent. The manufacturer can surpass the required local-content threshold by sourcing either the blades or one or more of the nacelle subcomponents in Ontario. Therefore, it is not necessary to assemble nacelles or produce blades locally to meet local content requirements. Vestas, for example, appears to be reaching the required domestic content level for wind farms in Ontario by sourcing nacelle components and controllers from Ontario rather than by producing locally. This does not mean that local content requirements have no impact on trade and competition in the market. However, it does indicate that nacelles and/or blades can be sourced from foreign locations.

Sources: Romano, "GE and Vestas Slam Canadian Provinces," October 20, 2011; North American Windpower, "Vestas Nacelles Signs Deal," January 6, 2012; Ontario Power Authority, "Feed-in Tariff Contract," July 15, 2011, 84–89; Ontario Power Authority, "Feed-in Tariff Program: FIT Rules," July 15, 2011, 16–17; Automodular Corp., "Annual Information Form," March 9, 2012, 7–8, 13–14; Wong and Zlahtic, "Insight into the Current Status," October 5, 2010, 8–9.

Table 5. U.S. nacelle producers active in Canada and their strategies for supplying the Canadian market, April 2012

OEM	2007–11 (MW)	Estimated pipeline (MW)	Canada	Other	United States
Acciona	45	431	No	Yes	Supplied nacelles for previous installation in Canada from
					U.S. plant; plans to export nacelles for other projects from the United States.

Table 5. (Continued)

OEM	2007–11 (MW)	Estimated pipeline (MW)	Canada	Other	United States
DSME/DeWind	0	10	No	No	Plans to export nacelles for this project from its plant in the United States.
Gamesa	0	130	No	Yes	Plans to export nacelles for 10 MW project in Ontario from the United States.
GE	1,078	1,284	Yes	Yes	Will likely meet demand in Quebec via local nacelle production. Since 2006, has likely met a portion of demand in Ontario via U.S. production and most demand in Alberta via U.S. production.
Northern Power Systems	0.3	0	No	No	Exports from the United States to Canada; all its production is domestic.
Siemens	952	720	No	Yes	Had limited or no exports before 2012, but plans to supply the majority of its U.S. and Canadian projects from its U.S. plant.
Vestas	1,216	992	No	Yes	Had limited or no exports to Canada before 2012, but plans to export products from its U.S. nacelle plant to Canada.
					Vestas nacelles for at least three orders scheduled for delivery in 2012—a 99 MW order, a 104 MW order, and a142 MW order—will be built in the United States.

Sources: Vestas, "Vestas Receives 104 MW Order," April 20, 2011; industry representative, interview by USITC staff, September 7, 2010; Vestas, "Vestas Bolsters Colorado Economy," July 13, 2010; Vestas, "Vestas Receives 99 MW Order," December 21, 2010; Stephens and Green, "Hutch Can Soar," December 4, 2010; industry representative, e-mail to USITC staff, October 1, 2010; USITC DataWeb/USDOC (accessed April 10, 2012); CANWEA Web site, http://www.canwea.ca/farms-farmse.php (accessed September 7, 2011 and April

2, 2012); Waggoner, "Achieving Supply Chain Success," June 5, 2012; Quilter, "Gamesa Wins First Canadian Order," February 20, 2012; AWEA, "Gamesa Turbines from Pa. Plant," February 23, 2012; North American Windpower, "DeWind to Supply Turbines to Research Site," May 26, 2011; Northern Power Systems Web site, http://www.northernpower.com/ about/index.php (accessed April 10, 2012); Capital Power Web site, http://www.capital power. com/community/ consultation engagement/Pages/ qualitywindproject.aspx and http://www. capitalpower.com /community/ consultationengagement /Documents /Quality-Wind/2012-spring-quality-wind-info- sheet.pdf (accessed July 11, 2012); David, Wind Turbines, June 2009, 20; Vestas Web site, http://www.vestas.com (accessed April 30, 2012); DeWind Web site, http://www.dewindco.com/eng/ (accessed April 30, 2012). See figures 6 and 8 for project pipeline sources and methodology.

U.S. producers have likely exported wind turbine blades to Canada in the past, though precise data are not available. Canadian imports of "parts of windmills" from the United States, which could include blades, hubs, and nacelle components, were less than $5 million annually during 2007–08, but averaged $50 million during 2009–11.[59] U.S. exports in the Schedule B provision that includes blades were also significantly higher during 2009–11 than in the prior two years.[60]

There are further export opportunities for blades, but it is not clear whether exports will be widespread. As with nacelles, Vestas invested in U.S. production in Colorado with the intention of also serving the Canadian market from these plants, and it has indicated that it plans to export blades for at least three projects totaling 345 MW from its U.S. plants.[61] Beyond this, however, little information is available on U.S. export potential. LM Wind Power is a supplier to multiple OEMs that have contracts in the Canadian market, but also has a plant in Canada and is likely to supply many Canadian projects from this plant, including the REpower projects in Quebec, where there is a local-content requirement.[62]

Siemens has a blade plant in Iowa, and has indicated that with the potential PTC expiration it may source more blades for the Canadian market from the United States, but is also building a blade plant in Ontario in response to the local content requirements.[63]

Information on whether other firms will source from U.S. suppliers or build blades at in-house U.S. plants for installations in Canada is not available.

Mexican Market Provides Opportunities for U.S. Producers

Overview

The Mexican market for wind power is smaller than the Canadian market, but it is growing rapidly, and there is no local production of nacelles. This rapid growth and lack of domestic competition suggest potential export opportunities for U.S. nacelle and blade producers. Although the firms that are supplying the Mexican market are mostly EU-based, these firms also have the option of exporting to Mexico from the United States, as they have U.S. nacelle and, in some cases, blade plants. These EU- based firms have generally chosen not to source from the United States in the past, but are likely to increase U.S. sourcing as the Mexican market expands and shifts toward the larger turbines commonly produced in the United States. In addition, some EU-based firms started U.S. production in 2010 and are more likely to export now that these firms have reached higher production rates.

Market

The wind turbine market in Mexico rapidly expanded during 2007–11 and is projected to grow further in the next few years. Annual wind turbine installations in Mexico increased from 0 MW in 2007 to 316 MW in 2010. Installations totaled only 50 MW in 2011, but more than 450 MW was completed in the first quarter of 2012.[64] Navigant's BTM Consult projects that annual installations in Mexico will total 600 MW in 2013 and 750 MW in 2015.[65] As of the end of 2012, most wind installations were in Oaxaca, which has some of the best wind resources in the world. A significant number of projects are also planned in Baja California, from which electricity can be exported to California.[66]

The expansion of the wind market in Mexico, which has excellent wind resources, has been driven by added transmission capacity, legislative and regulatory changes, declining wind turbine prices, and improving access to capital.[67] There are several regulatory or legislative changes that have played a role in the expansion of the wind market. In 1992, the Law of Public Service of Electricity was amended to open up power generation to the private sector, which is developing most new wind projects. The private sector generally participates in wind development in one of the following ways: (1) a private project developer bids on and wins a 25-year power purchase agreement (PPA) with the Federal Electricity Commission (CFE, the state-owned national utility) for the purchase of electricity; (2) self-supply arrangements, whereby an end-user becomes a part owner of the wind project and contracts to purchase the electricity generated for a 15- to 20-year period; or (3) a developer puts together a project that will export electricity. Producers of less than 30 MW can also sell their electricity to CFE, but pricing for this is not currently competitive.[68]

Later legislative or regulatory changes that benefited wind development made wind equipment eligible for accelerated depreciation, reduced electricity transmission charges, and created an “energy bank” in which self-supply customers can essentially store credit for any energy generated in excess of their use at a particular point in time.[69] In 2008, the Law for the Use of Renewable Energy and Financing of the Energy Transition (LAERFTE) set a goal of increasing renewables (excluding hydroelectric) from 3.3 percent to 7.6 percent of generation capacity and expanding them from 3.9 percent to 6.6 percent of actual generation. The law also addressed issues such as grid access, interconnection standards, and research and development.[70]

These policies have encouraged more wind energy development in Mexico, but they have not addressed all regulatory issues. Moreover, transmission capacity remains insufficient (despite the additional capacity brought online), and there can be significant permitting and land-leasing challenges in project development.[71] The United Nations Clean Development Mechanism, which allows facilities to generate clean carbon emissions certificates for global trading, remains an important component of making wind projects in Mexico economically viable.[72]

Financing can also be a challenge for projects in Mexico, particularly for small project developers, and any concerns that arise about the credit-worthiness of electricity purchasers for self-supply projects can make these projects harder to finance.[73] Project development in Mexico is currently being done by several different types of firms: Mexico-based project developers; at least one U.S.-based developer (Cannon Power Group); large international—especially Spain-based—developers (e.g., Iberdrola, Renovalia, and EDF Energies); and the development arm of some OEMs (e.g., Acciona and Gamesa).[74] Many of the projects, including those developed by large international project developers, involve a multitude of funders. These funders may include international development banks (e.g., the Inter-American Development Bank and North American Development Bank) and export credit agencies (e.g., Denmark's export credit agency, Eksport Kredit Fonden [EKF], and the Ex-Im Bank).[75]

Market Competition

Three companies, EU-based Acciona and Gamesa and U.S-based Clipper, accounted for all wind turbine installations in Mexico during 2009–11, but there is increasing competition in the wind turbine market (figure 9).[76] Gamesa has the largest project pipeline in Mexico, but GE, Siemens, and Vestas are increasing their presence in the market.[77] In addition, Asian wind turbine manufacturers are actively seeking to enter the Mexican wind turbine market.[78]

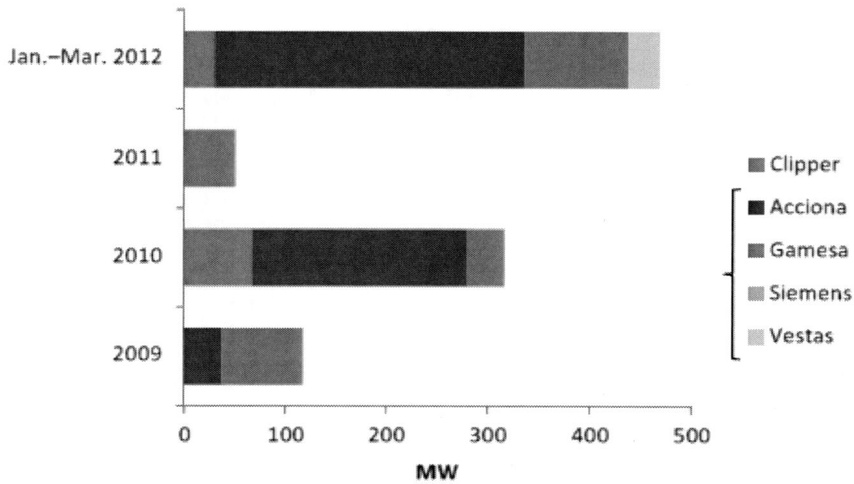

Source: AMDEE, "Wind Energy Projects," n.d. (accessed May 14, 2012); North American Windpower, "Iberdrola Commissions," January 17, 2012; Windpower Intelligence, "Last Clipper Windpower Turbine," March 9, 2012; North American Windpower, "Vestas Receives Order for Wind Project," July 5, 2011; Windpower Intelligence, "Mexico: Acciona," March 8, 2012.

Note: No projects were completed in 2007 and 2008.

Figure 9. EU-based companies are the leading suppliers to the Mexican market, as of March 2012.

All wind turbine demand is currently met from outside of Mexico; there is no domestic production of nacelles, and the only blade producer (Vientek, a joint venture of TPI Composites and Mitsubishi Heavy Industries) makes products for Mitsubishi, which does not have any recent Mexican installations.[79] Mexico's imports of wind-powered generating sets totaled $341.4 million in 2011, making it the ninth-largest global importer.[80]

U.S. Export Opportunities

The United States exported more than $50 million worth of wind-powered generating sets to Mexico during 2007–11 and was the second largest supplier to Mexico. Overall, however, the EU was the dominant exporter to Mexico during 2007–11 (figure 10). U.S.-based Clipper, which produces nacelles only in the United States, was the leading U.S. supplier to the market in Mexico, likely accounting for 59 of the 76 nacelles (78 percent) exported from the United States to Mexico during 2009–March 2012.[81]

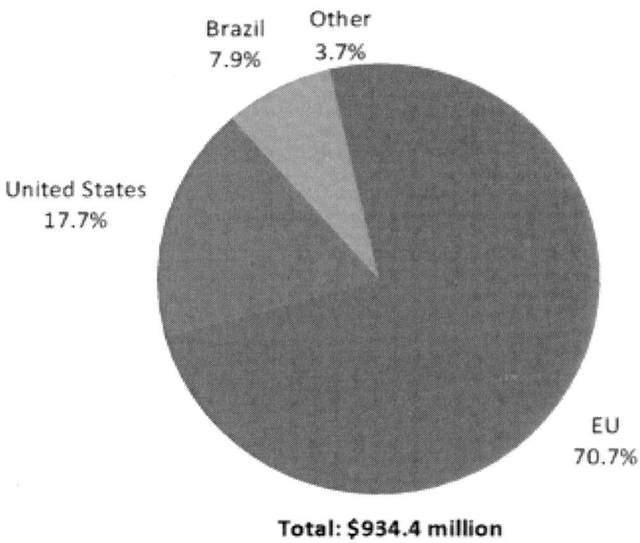

Source: GTIS, Global Trade Atlas database (March 21, 2012).
Note: Based on Mexican import data, which may differ from U.S. export data.

Figure 10.The EU was the leading source of Mexican wind-powered generating set imports during 2007–11.

The leading EU-based suppliers to the Mexican market have both U.S. and foreign production locations, but up to this point most have principally supplied the market from their nacelle plants in Europe.[82]

European firms have primarily served the Mexican market from their nacelle plants in Europe for several reasons. First, most imports from Denmark were in 2010, when the Siemens and Vestas plants in the United States were just opening and beginning to ramp up production.[83] Second, much of the demand in Mexico through the end of 2011 was for Gamesa's 850 kW turbines, which are not generally produced in the United States. Of the 246 Gamesa turbines installed during 2009–February 2012, all but 5 were 850 kW turbines.[84] Third, EU and European national export credit agencies have been a significant source of funding for projects in Mexico, but these may require the sourcing of equipment or services from the EU or the particular country where the agency is based. For example, Mexico's recently completed La Venta project, which received funding from the European Investment Bank (EIB), used 121 Gamesa 850 kW wind turbines from Spain.[85] Fourth, the prevalence of Spain-based project developers in the market may have influenced sourcing patterns. Spain-based project developers have primarily sourced wind turbines from the EU to this point, though whether this is due to a preference for these products or the other factors discussed here is not clear.[86] Finally, firms' decisions about allocating plant capacity across their global production network and relative levels of demand in the United States, Europe, and other markets may influence their sourcing decisions.

There is now more potential for EU-based firms to export from the United States to Mexico. Much of Gamesa's project pipeline, which is the largest in Mexico at about 948 MW, is for 2 MW wind turbines that can be made in the United States, and the company is planning to supply at least one project from its U.S. plant (table 6).[87] Vestas, which has the second largest pipeline in Mexico, has ramped up production at its plant in the United States and started exporting from the United States to Mexico in 2011.[88] GE also is planning to supply a project in Mexico, and given the proximity of the project to GE's U.S. plants, it is likely that GE will choose to supply this project from the United States. In addition, the project pipelines listed here may understate the potential for U.S. exports as many projects in development have not publicly announced a wind turbine supplier. Moreover, the listed pipelines only take into account the first phase of some projects that developers intend to ultimately be much larger. For example, Gamesa's approximately 948 MW project pipeline only includes the first (72 MW) phase of the Aubanel wind project being developed by Cannon Power Group, but the project could reach 1,000 MW in size.[89]

Table 6. U.S. nacelle producers active in Mexico and reported strategy for supplying the Mexican market, March 2012

Installations				Non-U.S. nacelle plants	Extent to which supply the Mexican market 2007– from the United States
2007– from the United States March 2012 (MW)		Under Construction	Estimated Pipeline (MW)		
Acciona	556	0	0	Yes	Limited or no exports in the past. Has indicated that it has international customers that may supply from U.S. plant, but no known projects under construction or in development in Mexico.
Clipper	148	0	0	No	Only produces in United States, so exports from Iowa plant for all installations.
Gamesa	218	164	948	Yes	Limited exports in the past as most turbines supplied have been 850 kW turbines. Plan to export nacelles for project in Baja California from the United States. Substantial portion of current project pipeline is for 2 MW turbines.
GE	0	0	253	Yes	Not available.
Siemens	0	161	99	Yes	Not available.
Vestas	29	102	549	Yes	Likely exported to Mexico in 2011 and 2012. Has indicated that it plans to export products from U.S. nacelle plant to Mexico.

Sources: Norfleet, "IWEA Head Withdraws 'Layoff' Quote," March 2, 2012; Vestas, "Vestas Bolsters Colorado Economy," July 13, 2010; industry representative, e-mail to USITC staff, October 1, 2010; Soto, "Del Mar Firm Signs Deal," May 20, 2010; USITC DataWeb/USDOC (accessed April 10, 2012); North American Windpower, "Iberdrola Commissions," January 17, 2012; North American Windpower, "Gamesa, Eolia Renovables De Inversiones," December 2, 2010; United Nations Framework Convention on Climate Change (UNFCCC) Clean

Development Mechanism (CDM) Web site, http://cdm.unfccc.int (accessed April 30, 2012); Biller, Calderón Inaugurates," January 23, 2009; North American Windpower, "Iberdrola Commissions," January 17, 2012; Iberdrola, "Iberdrola Renovables Buys," January 11, 2011; David, Wind Turbines, June 2009, 20; Vestas Web site, http://www.vestas.com (accessed April 30, 2012); GTIS, Global Trade Atlas database (accessed July 19, 2012). Project pipeline from AMDEE, "Wind Energy Projects," n.d. (accessed May 14, 2012), and other projects identified by USITC staff.

Notes: There are no nacelle plants in Mexico. The estimated project pipeline and project construction data here reflect a conservative estimate that includes only projects listed by the Mexican Wind Energy Association with secured access to transmission capacity and a small number of additional projects identified by USITC staff.

The growth in the wind turbine market in Mexico will result in some export opportunities for U.S. blade producers.[90] Gamesa plans to export blades from its U.S. plant for the 72 MW Aubanel wind project in Baja California and could potentially export to other Mexican projects from this plant.[91] Vestas has also indicated that it plans to export blades from its U.S. plant to projects in Mexico.[92] Other export prospects are uncertain. Acciona opened a blade plant in Spain in 2010 and sources many of the blades it uses in Mexico from this plant.[93] LM Wind Power supplies OEMs active in the Mexican market, but whether any of these firms will choose to source from LM Wind Power's U.S. blade plants is unknown. Siemens has U.S. blade production, but it is not known whether Siemens plans to source from the United States for any projects in Mexico. Finally, GE has multiple U.S. blade suppliers, but no information is available on whether it plans to use these suppliers for projects in Mexico.

Vibrant Brazilian Market Offers Inconsistent U.S. Export Opportunities

Overview

Brazil is by far the largest market for wind power in Latin America, and also has the largest potential for wind power in the region. Brazilian government policies favoring renewable energy and the country's eagerness to develop wind power, along with domestic content requirements, are leading most global wind turbine OEMs to locate production in the country. Some of these OEMs indicate that their Brazilian production facilities will become bases for exports to other Latin American countries. U.S. exports of wind turbines to Brazil are increasing at present, but as wind turbine OEMs increase their production operations in Brazil, future U.S. export potential may be limited.

Market

Brazil has significant potential for wind power, with the best resources being the Northeast, the Southeast, and the Southern Regions. Brazil's cumulative installed wind power capacity rose from 247 MW in 2007 to 1.5 GW in 2011.[94] Since Brazil's potential wind energy is estimated at 305 GW, there is considerable scope for expansion.[95]

In recent years, the growth in Brazil's renewable energy market was principally driven by the PROINFA program (Programa de Incentivo às Fontes Alternativas de Energia Elétrica). Established in 2004 and terminated in 2011, PROINFA was designed to develop renewable energy sources for Brazil's electricity portfolio.[96]

The program used a feed-in tariff (FIT) scheme requiring the state-owned electricity company Electrobrás to buy electricity from renewable energy producers under 20-year power purchase agreements at rates higher than those of conventional power sources. At the end of 2011, the PROINFA program had developed 1,326 MW of wind power.[97]

To add further renewable energy production capacity, in 2008, the Brazilian government launched a policy whereby electricity generating companies must bid to sell their electricity into the national grid system. This is done through auctions that cover wind, hydro, biomass, and other power generation sources.

Brazilian Electricity Regulatory Agency (ANEEL) establishes the regulations for the auctions, which are conducted by a nonprofit organization, the Chamber of Electrical Energy Commercialization. Those companies offering the lowest price per megawatt-hour (MWh) win the bidding. Six auctions for wind power were held between December 2009 and December 2011, resulting in a total of 6,876 MW under contract.[98]

Brazil also has an increasing amount of wind power development in the non-regulated market, called the Ambiente de Contratação Livre (ACL), or the Free Contract Environment, in which electricity prices are freely negotiated under power purchase agreements between wind power producers and electricity consumers. Wind installations in this market segment totaled at least 14 MW in 2011, with another 546 MW under development.[99]

Many of Brazil's wind energy projects have obtained long-term project financing from the government-owned National Economic and Social Development Bank (Banco Nacional de Desenvolvimentao Económico e Social, or BNDES).[100]

Credit approvals for wind energy by BNDES totaled approximately $140 million in 2008; $601 million in 2009; $682 million in 2010; and $2,033 million in 2011.[101]

Some of the projects financed by BNDES occurred under the Brazilian Growth Acceleration Program (PAC), begun in January 2007.[102] PAC is a major Brazilian economic stimulus initiative, which included approximately $150 billion for energy infrastructure development.

In addition, since 1997, the wind energy industry in Brazil has benefited from exemptions from the *Imposto sobre Circulação de Mercadorias e Serviços* (ICMS), the merchandise and service circulation tax, which is a value-added tax on imports and domestically produced goods.[103] In December 2010, the exemption was expanded to include wind turbine blades; previously only nacelles and towers were covered.[104] Without the exemptions, the ICMS on wind turbine components would be approximately 17 percent of their value. ICMS exemptions are issued for a 2-year period, and have been consistently renewed.

The latest exemption agreement for the wind power industry was renewed in January 2010 and will expire in December 2012.

Another significant tax exemption affecting the Brazilian wind turbine market involves the *Imposto sobre Productos Industrializados* (IPI), which is a tax on imported and domestically produced manufactured goods. In December 2006, the Ministry of the Treasury exempted parts of wind turbines from the IPI.

In June 2009, that exemption was eliminated, but in December 2009, the Ministry of the Treasury, as part of a national economic stimulus plan, again exempted parts of wind turbines from the tax, which was approximately 7.5 percent of the value of the equipment.[105]

Market Competition

The Brazilian wind turbine and component manufacturing industry is rapidly expanding due to the recent influx of foreign investment in the sector.

Table 7. Principal wind turbine and component manufacturing operations in Brazil, and their status, March 2012

Manufacturer	Location (city, state)	Product	Date announced	Date operational
Acciona (Spain)	Unknown	Hubs and nacelles	April 2012 (nacelles), July 2012 (hubs)	October 2012 (hubs), no date set for nacelles
Aeris Energy/Suzlon (India)	Pecém, Ceará	Blades	February 2011	August 2011
Alstom Power (France)	Camaçari, Bahia	Nacelles	December 2009	November 2011
Fuhrländer AG (Germany)	Pecém, Ceará	Nacelles	August 2009	N/A
Gamesa (Spain)	Camaçari, Bahia	Nacelles	October 2010	July 2011
GE Energy (United States)	Campinas, São Paulo	Wind turbines	November 2009	December 2009
GE Energy (United States)	Bahia	Wind turbines	September 2011	Mid-2013
IMPSA (Argentina)	Suape, Pernambuco	Wind turbines	N/A	September 2008
IMPSA (Argentina)	Rio Grande do Sul	Wind turbines	April 2012	2013
LM Wind Power	Suape, Pernambuco	Blades	December 2011	June 2013
Renobrax Energias Renováveis	Guaíba, Rio Grande do	Wind turbines	October 2009	October 2011
(Brazil)	Sul			
Siemens (Germany)	Unknown	Unknown	2009	N/A
Sinovel (China)	Unknown	Wind turbines	September 2011	N/A
Tecsis (Brazil)	Sorocaba, São Paulo	Blades	N/A	1995
Vestas (Denmark)	Maracanaú, Ceará	Nacelles	August 2011	May 2012
Wobben Windpower (Enercon GmbH—Germany)	Sorocaba, São Paulo	Nacelles	N/A	1995
Wobben Windpower (Enercon GmbH—Germany)	Pecém in Ceará	Blades	N/A	2002

Source: Compiled from news releases, media reports, company Web sites and documents, and Bloomberg New Energy Finance database (accessed April 16, 2012).

Note: N/A: not available. The term "wind turbine" is used where more specific information is not available, but in most cases these plants likely produce nacelles. Siemens has not made a firm decision on whether to proceed with building a plant in Brazil.

Before 2008, Brazil had two wind turbine producers, Wobben Windpower Ind. e Com. and IMPSA S.A.—both subsidiaries of foreign producers—as well as indigenous production of blades and castings (table 7). The vast majority of wind turbine OEMs opened production in 2011 or have openings planned for 2012 and 2013.

These companies are establishing wind turbine assembly facilities in Brazil because of (1) the large size of the wind power market and the dramatic rise in wind power development in Brazil and other Latin American countries; (2) Brazilian domestic content requirements under the PROFINA program; (3) BNDES requirements for obtaining financing (discussed later); and (4) incentives from Brazilian states for foreign investment.

The Brazilian wind turbine industry is large enough to have begun exporting some wind-powered generating sets (HS 8502.31). Exports rose from $1 million in 2007 to $25.9 million in 2010, but totaled only $564,000 in 2011 as producers focused on fulfilling domestic orders.[106] During 2007–11, primary destinations for Brazilian exports of wind-powered generating sets were Costa Rica ($12.8 million in 2008 and $10.9 million in 2009), the Netherlands Antilles ($4.9 million in 2009), and Argentina ($25.8 million in 2010).

OEM wind turbine producers Gamesa and Alstom plan to export from their Brazilian factories, as does blade producer LM Wind Power.[107] Brazilian blade producer Tecsis was an early entrant to the U.S. and other markets and has been producing blades for over 10 years.[108]

U.S. EXPORT OPPORTUNITIES

Brazilian imports of wind-powered generating sets have risen steadily in tandem with the development of wind power in the country (table 8 and figure 11). In 2011, Germany became the leading supplier to Brazil, followed by the United States, Spain and India.

The rise in imports since 2009 corresponds with the first government auctions for wind power. Chinese firms are increasing their presence in the market, with Sinovel scheduled to provide 23 wind turbines to a Brazilian wind energy project by July 2012. The wind energy project is receiving funding from the China Development Bank.[109]

Table 8. Brazilian imports of wind-powered generating sets, 2007–11, million $

	2007	2008	2009	2010	2011
Germany	0.4	0.4	0.0	40.7	133.2
United States	14.9	14.1	8.7	5.9	109.7
Spain	0.0	0.0	0.0	84.2	93.0
India	27.0	107.2	160.1	134.0	64.0
Denmark	0.0	0.0	0.0	0.0	56.1
China	0.0	0.0	48.5	9.2	0.1
All others	0.0	0.1	3.8	0.0	0.1
Total	42.3	121.7	221.1	274.0	456.3

Source: GTIS, Global Trade Atlas database (accessed May 1, 2012).

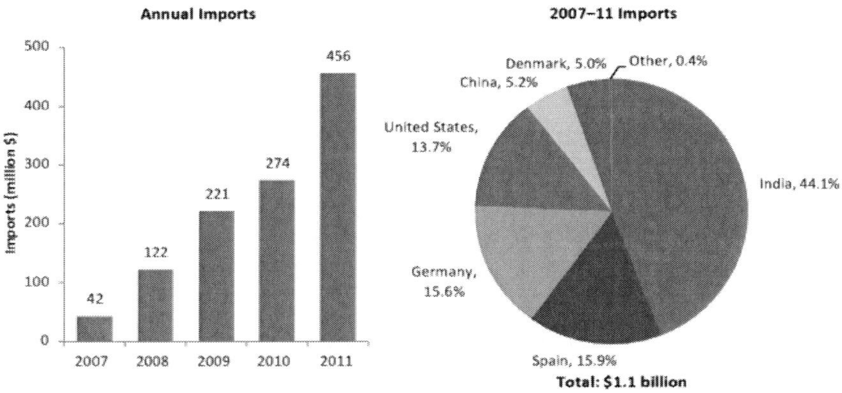

Source: GTIS, Global Trade Atlas database (accessed May 1, 2012).
Note: Based on Brazilian import data, which may differ from U.S. export data.

Figure 11. Brazilian imports of wind-powered generating sets are rapidly increasing.

U.S. producers have had some success in exporting to Brazil since the start of 2011. U.S. exports of wind-powered generating sets to Brazil, likely by GE, totaled $152.3 million from 2010 through the first quarter of 2011.[110] In 2012, LM Wind exported 60 wind turbine blades from its plant in North Dakota to Brazil and secured Ex-Im Bank financing for the export of blades from its plant in Arkansas to Brazil.[111]

There are, however, trade barriers that may limit future U.S. exports to Brazil. Brazil has implemented domestic content and import tariff policies to develop the Brazilian wind energy equipment manufacturing and service industry. Under the PROINFA program, wind farms selected for the program were required to have a minimum domestic content of at least 60 percent of the value of equipment and services used in the project.[112] Also, BNDES has the same requirement for its wind energy loans.[113] As a significant number of the wind energy projects under PROINFA and the ANEEL auctions have obtained BNDES financing, there has been a de facto minimum 60 percent domestic content-level in Brazilian wind energy projects.

The relatively high levels of domestic content in wind energy projects are also required in financing provided for projects in Northeast Brazil by the Bank of Northeast Brazil (BNB) and the Development Agency of the Northeast (ADENE). The use of domestically produced wind turbines also shows up in the requirements for the auctions under ANEEL for electricity supplied to the national grid. The Ministry of Mines and Energy initially required that wind energy projects could only use imported wind turbines with a capacity of over 2MW, but that threshold was reduced to 1.5 MW.[114] Regulations for subsequent auctions did not specify requirements regarding imported wind turbines— but if financing is from BNDES, then a 60 percent domestic content level is still needed.

BNDES conducted an audit in June 2012 of wind turbine suppliers, and reportedly suspended the distribution of loans to projects using turbines from several OEMs, including Acciona, Fuhrländer, Suzlon, and Vestas, because they did not meet local content requirements. This affects approximately 2 GW of projects for which these companies are suppliers.[115]

Brazil's import tariffs on wind turbines have fluctuated. The import tariff on wind turbines was lowered from 14 percent to zero in March 2007 to assist wind energy projects in meeting timely completion deadlines under the PROINFA program.[116]

But in June 2009, the Council on Foreign Trade (Câmara de Comércio Exterior (CAMEX)) raised the import tariff on wind turbines with a capacity of 3.3 megavoltamperes (MVA) and below from free to 14 percent. This change went into effect on January 1, 2010.[117]

U.S. export opportunities to Brazil exist for global OEM producers located in the United States, but these may be limited due to a significant number of wind turbine assembly facilities beginning operations in Brazil during 2011-13.

However, U.S. exports to Brazil may fill production constraints by these global OEM producers at their Brazilian factories, or supply demand from the non-regulated market for wind power.

AS OTHER LATIN AMERICAN MARKETS EXPAND, EXPORT OPPORTUNITIES ARE INCREASING

Overview

Other markets in Latin America are also expanding, and many markets could offer at least limited export opportunities, though these opportunities will vary by country. Firms with U.S. production have significant project pipelines in the region and may choose to supply some of these projects from their U.S. plants. U.S. producers are currently supplying the Central American market, planning to export to Uruguay, and have exported to Chile in the past. The extent of U.S. export opportunities will be affected by factors such as whether global OEMs export a significant share of nacelles for these projects from Brazil, how they choose to allocate production across their global manufacturing plants, and increasing competition from local and Chinese firms.

Market

The market in Latin America (excluding Brazil and Mexico) is currently small compared to the global market, with less than 1 GW in cumulative installations (table 9). In 2011, only 270 MW was installed in Latin America (excluding Brazil and Mexico), accounting for less than 1 percent of 2011 global installations.[118] However, substantial wind development is taking place in the region and Navigant's BTM Consult projects that annual installations in Latin America (excluding Brazil and Mexico) will increase from 270 MW in 2011 to 750 MW in 2014 and 1,100 MW in 2016.[119]

The main drivers of the wind market in Latin America (excluding Brazil and Mexico), though they vary by country, are:

- New policies to encourage the development of wind power;
- Rising electricity demand;
- Declining wind turbines prices;
- Lower hydroelectric production in recent dry years and resulting higher electricity prices;

- The need to complement hydropower with another power generation source during the dry season, along with the fact that wind generation in some areas peaks when hydro generation is at its lowest;
- The cost-competitiveness of wind-generated electricity in some countries;
- Excellent wind resources;
- A need to reduce dependence on imported fossil fuels and mitigate the risk of fossil fuel price fluctuations; and
- A desire to install energy generation technologies with a low environmental impact and/or that are carbon free.[120]

Table 9. Latin American annual installed wind energy capacity, MW, 2007–11 (excluding Brazil and Mexico)

	2007	*2008*	*2009*	*2010*	*2011*
Honduras	0	0	0	0	102
Argentina	2.0	0	0	25.3	79
Chile	18.1	0	147.5	2.6	33
Dominican Republic	0	0.2	0	0	33
Costa Rica	0	0	49.0	0	13
Nicaragua	0	0	40.0	0	N/A
Uruguay	0.4	19.9	0	10.0	N/A
Jamaica	0	0	34.0	0	0
Netherlands Antilles	0.3	0	12.0	0	N/A
Colombia	0	0	0.5	0	N/A
Cuba	1.6	5.1	0	4.5	0
Ecuador	3.1	0.9	0	0	N/A
Dominica	0	0.2	0	0	0
Other/not specified	0	0	0	0	10
Total	25.5	26.3	283.0	42.4	270

Sources: WWEA, World Wind Energy Report 2010, April 2011, 19–20; WWEA, World Wind Energy Report 2009, March 2010, 16–17; GWEC, Global Wind Report: Annual Market Update 2011, March 2012, 11.

Note: N/A: not available. Only includes countries with installations during 2007–11. Data from the Bahamas, the British Virgin Islands, El Salvador, French Guiana, Guatemala, Panama, Paraguay, Suriname, and Venezuela are not reported by the World Wind Energy Association. Data for 2011 come from GWEC. The "other" category in the 2011 data includes Colombia, Ecuador, Nicaragua, Peru, and Uruguay. GWEC data are rounded to the nearest MW.

Some countries also have an interest in developing wind projects to export electricity.[121] Despite this increasingly favorable environment for wind, a number of significant barriers to wind development remain including:

- The poor electric grid infrastructure;
- A shortage of skilled personnel with renewable energy experience;
- Utilities' lack of familiarity with wind;
- Challenges financing projects; and
- Continuing insufficient policy support, despite recent improvements.[122]

MARKET COMPETITION

EU-based companies were the dominant suppliers to Latin American projects (excluding Brazil and Mexico) during 2007–11 (figure 12). Vestas alone accounted for more than 40 percent of projects installed during 2007–11 and is widely present in the region, having supplied turbines to at least 10 countries during its history.[123] Gamesa was second in terms of MW installed, mostly due to a 102 MW project recently completed in Honduras.[124]

There are several local wind turbine producers that are currently competing against foreign suppliers in these Latin American markets. Argentina has three wind turbine manufacturers, IMPSA, Patagonia NRG, and INVAP (table 10). IMPSA produces 1.5 MW, 1.8 MW, and 2.1 MW wind turbines and is also active in project development.[125] With 4 percent of installations during 2007–11 and a significant project pipeline, as discussed below, IMPSA is a major supplier in the region.[126] IMPSA recently indicated that it is increasing its focus on Latin American markets other than Brazil, including Venezuela, Argentina, and Uruguay, as a result of the intense competition in Brazil and consequent decline in prices and profits in the Brazilian market.[127] Patagonia NRG produces a 1.5 MW wind turbine and INVAP is developing a 1.5 MW turbine.[128]

Industria Tecnológica Panameña S.A (I.T.P) subsidiary Turbowinds produces utility-scale wind turbines in Panama. Turbowinds currently produces a 600 kW wind turbine and, according to I.T.P., it expected to have a 1.6 MW turbine by the end of 2011.[129] Chinese companies, facing overcapacity and a stagnant market at home, are also seeking sales in Latin America. Goldwind has already achieved a significant sales volume in the region. Goldwind's first nacelles shipped to Latin America were installed in a 4.5 MW project in Cuba in 2011 and it has contracts to supply another 121.5 MW to projects in Chile and Ecuador.[130] China also exported $41 million in wind-powered generating sets to Argentina in 2011 and XEMC has an upcoming 100 MW project in Argentina that will use wind turbines from China.[131]

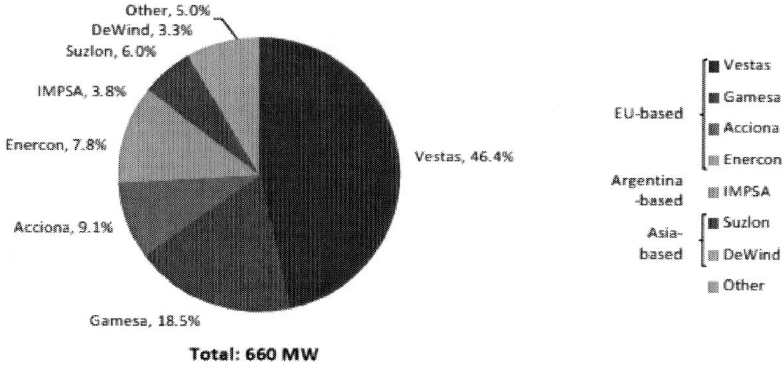

Source: Bloomberg New Energy Finance database (accessed March 22, 2012).
Note: Total may differ slightly from GWEC and WWEA data.

Figure 12. Vestas and other EU-based firms were the leading suppliers to Latin American markets (excluding Brazil and Mexico) during 2007–11.

Table 10. Latin American nacelle producers, excluding Brazil and Mexico, March 2012

Manufacturer	Product
Argentina	
IMPSA, SA	1.5 MW, 1.8 MW, and 2.1 MW turbines
INVAP	Developing a 1.5 MW turbine
NRG Patagonia	1.5 MW turbine
Panama	
I.T.P./Turbowinds	600 kW, 1.6 MW was expected to be developed by the end of 2011

Source: I.T.P. Web site, http://www.itp.com.pa/index.html (accessed April 17, 2012); Turbowinds Web site, http://www.turbowinds.com/index.php (accessed April 17, 2012); Patagonia NRG Web site, http://www.nrgpatagonia.com/ (accessed April 17, 2012); INVAP Web site, http://www.invap.com.ar (accessed April 17, 2012); IMPSA Web site, http://www.impsa.com (accessed April 17, 2012).

Note: The Turbowinds and I.T.P Web sites both list 600 kW wind turbines, but for the other turbines produced by the company the two Web sites list different turbine sizes. This table relies on the more recently updated I.T.P. Web site.

Other firms have not yet made major sales inroads into Latin America (excluding Brazil and Mexico). GE was selected to supply turbines for one upcoming 50 MW project, but otherwise U.S.-based firms do not have many sales in the region.[132] India-based Suzlon and Korea-based DeWind—which only produces in the United States—and Unison have also had a small volume of sales in the region.[133]

Vestas and Gamesa have the strongest project pipelines in Latin America (excluding Mexico and Brazil), with the highest number of projects, the greatest geographical diversity, and the largest total MW in the project pipeline (figure 13).

IMPSA, other EU-based firms, and China-based Goldwind have significant projects in development, but these firms are not supplying as many projects and are not present in as many markets.[134] When the pipelines of the other EU firms are taken together with the Gamesa and Vestas pipelines, however, it is apparent that EU-based firms and Argentina-based IMPSA currently account for the bulk of the market.

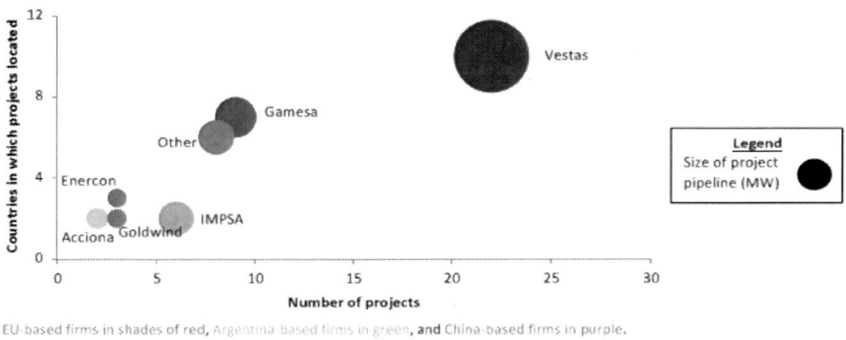

Source: North American Windpower, "Pattern Energy Developing 115 MW," November 28, 2011; United Nations Framework Convention on Climate Change (UNFCCC) Clean Development Mechanism (CDM) Web site, http://cdm.unfccc.int (accessed April 17, 2012); Bloomberg New Energy Finance database (accessed March 22, 2012); REVE, "La Rioja Signs Contract," November 2, 2010; IMPSA, "Uruguayan UTE Approves Contract," n.d.; IMPSA, Corporate Presentation, February 2012, 14.

Notes: Size of bubble represents number of MW in the project pipeline. Includes firms with either at least 150 MW or at least 3 projects.

Figure 13. Vestas and Gamesa have the strongest project pipelines in Latin America (excluding Brazil and Mexico), April 2012.

U.S. EXPORT OPPORTUNITIES

The United States is a significant supplier to Latin America (excluding Brazil and Mexico), shipping nacelles for almost 20 percent of installed capacity during 2007–11 and accounting for 11 percent of global exports to Latin America (figure 14). However, the U.S. share of the market is small compared with that of the EU, and Brazil and China also account for a larger share of exports to the region than the United States. The largest projects supplied by U.S. producers during 2007–11 were a 102 MW Gamesa project in Honduras and a 20 MW project in Chile supplied by DeWind.[135]

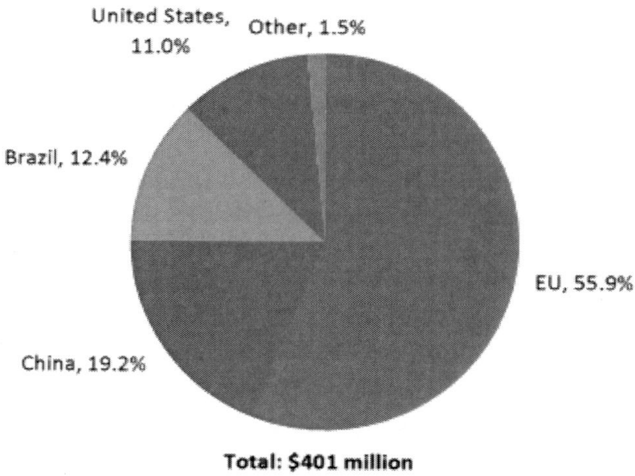

Source: GTIS, Global Trade Atlas database (accessed March 23, 2012).
Notes: Chart shows shares of global exports to Latin America, based on each country's exports under HS 8502.31. Data are from GTIS and may differ slightly from U.S. data found in DataWeb. Indian exports for 2011 are estimates based on partial-year data.

Figure 14. The EU was the leading exporter to Latin America (excluding Mexico and Brazil) during 2007–11.

Firms with U.S. nacelle production have a project pipeline in Latin America (excluding Mexico and Brazil) of more than 2.9 GW and could export from the United States for some of these projects (table 11). Of the seven companies with orders in these countries, at least three have exported to Latin America (excluding Brazil and Mexico) in the past and at least one other firm is considering future exports. The extent of U.S. export opportunities, however, is likely to vary by country.

Table 11. U.S. nacelle producers active in Latin American markets (excluding Brazil and Mexico) and reported strategy for supplying these markets, April 2012

Company	Estimated project pipeline (MW)	Non-U.S. nacelle plants		Extent to which supply the Latin American market
		Brazil	Other	(excluding Brazil and Mexico) from the United States
Acciona	153	Yes	Yes	No information on intent to supply from the United States.
DeWind	10	No	No	Exported from the United States to Chile in the past. Only produce nacelles in the United States.
Gamesa	586	Yes	Yes	Exported nacelles from the United States for 102 MW project in Honduras that received Ex-Im Bank funding.
				Plans to export from the United States for projects in Nicaragua and Uruguay, though has also indicated that may use new plant in Brazil to supply nearby countries.
GE	50	Yes	Yes	No information on intent to supply from the United States.
Nordex	130	No	Yes	Is considering options for exporting from the United States.
Siemens	115	Yes	Yes	No information on intent to supply from the United States.
Vestas	1,880	Yes	Yes	Likely exported from the United States to South America in the past and is producing in the United States for at least one project in Central America.
				Reportedly has cost advantages in supplying Latin American projects from the United States rather than from the EU.

Sources: Ex-Im Bank, "Export-Import Bank Provides $159 Million," August 20, 2010; Gamesa, "Gamesa Announces Industrial Project," December 1, 2010; Business News Americas, "In Brief: DeWind Shipping," December 31, 2008; North American Windpower, "Pattern Energy Developing 115 MW," November 28, 2011; UNFCCC CDM Web site, http://cdm.unfccc.int (accessed April 17, 2012); David, Wind Turbines, June 2009, 20, 41–42; DeWind Web site, http://www.dewindco.com (accessed April 17, 2012); GTIS, Global Trade Atlas database (accessed July 11, 2012); Bloomberg New Energy Finance database (March 22, 2012); Northern Colorado Business Report, "Vestas Colo. Shipping," December

16, 2011; Gamesa, "Gamesa Strengthens its Presence," May 28, 2012; industry representative, interview by USITC staff, May 1, 2012. See Brazil section for list of firms with manufacturing plants in Brazil.

Notes: Total in this table may differ from total in next table due to rounding. Includes planned manufacturing plants in Brazil.

In South America, the three countries outside Brazil with the largest near-term markets are Argentina, Chile, and Uruguay. While Peru's market is small, both Gamesa and Vestas have been selected to supply turbines to upcoming projects, indicating at least some export potential.[136] Other South American countries also have only small wind markets, though there is discussion in some countries of increasing wind capacity. If there is significant wind development in Colombia, U.S. exports could benefit from the implementation of the U.S.-Colombia free trade agreement (FTA), which eliminates Colombia's 5 percent duties on nacelles and blades.[137] The export prospects for Argentina, Chile, and Uruguay are described below:

Argentina Argentina has a growing wind market; 750 MW in projects were awarded in its first wind tender (Genren I) and almost 1.1 GW were expected to be awarded in a second tender (Genren II), though financing problems have slowed the development of the first round of wind projects.[138] Significant competition in Argentina's wind market may limit U.S. exports prospects. Several global OEMs are planning to supply Argentina's markets from Brazil and Argentina's domestic producers have significant market share. In addition, Chinese firms appear to be making significant inroads into the market. China accounted for 69 percent of Argentina's imports of wind-powered generating sets in 2011, Chinese firm XEMC has a contract for 100 MW in Argentina, and the developer of a recently announced wind project of more than 1 GW is planning to use Chinese wind turbines. The availability of Chinese financing was a critical factor in the decision to source from China for this 1 GW wind project and is a competitive advantage of the Chinese industry in Argentina.[139] The only firm with U.S. nacelle production that was identified as having been selected for projects in Argentina was Vestas, which is providing turbines for more than 400 MW of projects (table 12).[140] Overall, only limited information is available on how global OEMs intend to supply this market, but there appear to be challenges for U.S. firms seeking to enter the market.[141]

Uruguay Uruguay has held several rounds of wind tenders and Brazil and Uruguay agreed to jointly develop a wind project near the Brazilian border.[142] At least three firms with U.S. nacelle production, Gamesa, Nordex, and

Vestas, have been selected by project developers to provide equipment for winning bids in Uruguay's wind tenders.[143] U.S. producers appear to have significant interest in exporting to Uruguay and are actively exploring producing in the United States for projects in Uruguay. Gamesa is planning to export for one project in Uruguay from the United States and Nordex is reportedly considering doing the same.[144]

Table 12. Tariffs in select Latin American markets and pipeline for U.S. producers, March 2012

	Project pipeline for firms with U.S. plants (MW)	*MFN duty rate (%)*	*Duty rate for U.S. goods (%)*	*MFN duty rate (%)*	*Duty rate for U.S.goods (%)*
Chile	903	6	0	6	0
Panama	594	10	0[a]	3.7	0a
Argentina	414	0	0	14	14
Peru	234	0	0	0	0
Uruguay	199	0	0	1.7	1.7
Dominican Republic	175	0	0	0	0
Nicaragua	126	0	0	0	0
Costa Rica	100	0	0	0	0
Venezuela	100	10	10	5	5
Curaçao	54	N/A	N/A	N/A	N/A
Guatemala	21	0	0	0	0
St. Kitts and Nevis	5	5	5	5	5
Colombia	0	5	0	5	0
Aruba	0	N/A	N/A	N/A	N/A
Ecuador	0	0	0	2.5	2.5
Barbados	0	5	5	5	5

Source: WTO Tariff Download Facility (accessed March 22, 2012 and April 17, 2012); text of U.S. FTAs; see project pipeline figure for project data sources.

Note: N/A: Not available. If there are multiple tariff lines, rate of duty is a simple average of the duties in those tariff lines. Total in this table may differ from total in prior table due to rounding. All tariff rates are for 2011 except Argentina (2010), the Dominican Republic (2008), Guatemala (2009), Nicaragua (2009), and Peru (2009).

[a] Upon implementation of the trade promotion agreement.

Like Argentina, however, some OEMs are also planning to serve the market from Brazil. In announcing its plant in Brazil, Gamesa indicated that it would export to Uruguay from Brazil, but it is pursuing Ex-Im Bank funding for exports to Uruguay and decided to supply its first project in Uruguay from the United States.[145] In Uruguay's first wind tender, one of the criteria for rating proposed projects was the share of domestic content, which benefited Argentina-based IMPSA because its proposal included producing concrete towers on-site, even though the turbines will be supplied from Brazil.[146] Nevertheless, a significant number of EU-based OEMs are supplying projects in Uruguay and these firms are exploring the possibility of exporting from the United States.

Chile Chile has extensive project development underway and could offer some export opportunities for U.S. nacelle producers. Chile's market is competitive, with EU- and Asia-based firms generating sales, though EU-based firms account for most of the project pipeline. Four firms with U.S. production have pending projects, for a combined project pipeline in Chile of more than 900 MW.[147] DeWind's activity in the Chilean market (past and pending projects totaling 30 MW) leads directly to U.S. exports, as DeWind only produces in the United States.[148] The three other firms with U.S. production, Acciona, Siemens, and Vestas, have been selected to provide turbines to Chile. Vestas has likely supplied wind turbines to Chile from its U.S. plant in the past, but is opening a plant in Brazil and could choose to source either from Brazil or from another global production location for projects in Chile.[149] Acciona and Siemens have not moved as quickly to produce in Brazil, but would still have the option of sourcing from non-U.S. production locations for these projects. While the U.S.-Chile FTA eliminates the 6 percent tariff on wind-powered generating sets, other countries also benefit from FTAs with Chile.[150]

Central America has substantial wind resources and could be a significant U.S. export market if the pace of wind development picks up. U.S. producers could benefit from their proximity and, in Panama, from the elimination of the duty on wind-powered generating sets upon implementation of the trade promotion agreement.[151] Gamesa exported wind turbines to Honduras in 2011, as noted earlier, and Acciona, Gamesa, and Vestas have been selected to supply wind turbines for projects in Costa Rica, Guatemala, Nicaragua, and/or Panama.[152] Firms with U.S. nacelle production have been selected to supply wind turbines for more than 800 MW of projects. While all of these nacelles may not necessarily be exported from the United States, Gamesa and Vestas plan to at least partially source from the United States for projects in Nicaragua.[153]

Markets in the Caribbean are close to the United States, likely providing transportation cost advantages for U.S. nacelle producers. These markets are likely to remain small, however. Also, at least one recent installation used turbines less than 1 MW in size, which are not as widely produced in the United States.[154] The largest potential market appears to be the Dominican Republic, where Gamesa, GE, Nordex, and Vestas—all of which have U.S. nacelle production—have been selected to provide wind turbines.[155] The Caribbean has excellent wind resources, but overall electricity demand is relatively low. Installed electric generating capacity in the Caribbean (excluding U.S. territories) is only about 17 GW, with about 63 percent of that generating capacity in Cuba and the Dominican Republic.[156] One compilation of estimates of wind potential in nine markets in the region (Antigua and Barbuda, Barbados, the Dominican Republic, Guadeloupe, Haiti, Jamaica, Saint Kitts and Nevis, Saint Vincent and the Grenadines, and Trinidad and Tobago) indicated that up to 3.8 GW of wind power could be installed, with 85 percent of that potential in the Dominican Republic and 11 percent in Antigua and Barbuda.[157] During 2007–11 there was one U.S. export to Saint Kitts and Nevis, and also one export of a turbine to the Bahamas by Northern Power Systems.[158]

In the past, U.S. producers have also exported blades to Latin American markets. For example, LM Wind Power in North Dakota produced blades for the Gamesa wind turbines that were installed in Honduras, and it is likely that blades for the DeWind turbines exported to Chile were also produced in the United States.[159] As with nacelles, competition from producers in Brazil and the potential for EU-based firms to supply blades from other production locations may limit U.S. exports.160 Overall, the country-by-country trends for blades are likely to be similar to those for nacelles.

CONCLUSION

U.S. exports of wind turbine nacelles and blades worldwide are limited, but are increasing and appear poised for further growth in the next few years. Many of the leading global OEMs, and their blade suppliers, now produce in the United States, and the nearby Canadian and Latin American markets are rapidly increasing in size. Markets in Ontario, Western Canada, and Mexico, in particular, appear to offer significant near-term export possibilities as they are dominated by U.S.- and EU-based firms, many of which have U.S. plants from which they intend to supply a share of the turbines sold in these markets.

Many of the other markets in Latin America are highly competitive, but Central America, the Caribbean, and certain South American markets like Chile and Uruguay could offer export opportunities, depending on how firms choose to allocate production across their global manufacturing plants.

Some growing markets, however, will provide only limited export opportunities. The wind market in Quebec is substantial, but offers no export opportunities due to local-content requirements and the fact that the two producers that won contracts in recent wind tenders do not produce nacelles or blades in the United States. Brazil offers some export opportunities for U.S. producers, but these may be limited due to local-content requirements and a significant number of wind turbine assembly facilities beginning operations during 2011–13 in Brazil. Once these plants are open, U.S. exports to Brazil will likely occur only when Brazilian plants are operating near capacity or firms receive orders for the non-regulated wind market.

Overall, the U.S. industry benefits from its proximity to markets in Canada and Latin America, and resulting lower transportation costs for nacelles and blades. This will likely lead to further exports in the next few years, but expectations should remain modest as these markets, though growing, remain small in global context. In addition, nontariff barriers may impact trade, and it remains unclear how OEMs will allocate production across their global supply chains.

REFERENCES

Acciona. "Acciona Opens 25m Euro Wind Turbine Blade Manufacturing Plant at Lumbier (Navarre, Spain)." News release, June 30, 2010. http://www.acciona.com/news/acciona-opens-25m-eurowind-turbine-blade-manufacturing-facility-at-lumbier-%28navarre,-spain%29-.

American Wind Energy Association (AWEA). *Annual Wind Industry Report: Year Ending 2008.* 2009. http://www.awea.org/learnabout/publications/upload/AWEA-Annual-Wind-Report-2009.pdf.

———. *U.S. Wind Industry Fourth Quarter 2011 Market Report.* January 2012. http://www.awea.org/learnabout/publications/reports/upload/4Q-2011-AWEA-Public-MarketReport 1-31.pdf.

———. *U.S. Wind Industry Year-End 2010 Market Report*. January 2011. http://www.awea.org/learnabout/publications/upload/4Q10-market-outlook-public.pdf.

———. *Year-End 2009 Market Report*. January 2010. http://www.awea.org/learnabout/publications/upload/4Q09.pdf.

———. "Gamesa Turbines from Pa. Plant Headed for Canada." Blog, February 23, 2012. http://www.awea.org/blog/index.cfm? customeldataPageID1699 =14323.

Araluce y Martinez de Avagra, Juan. "Clean Energy to Keep Latin America Competitive." *World Energy Insight*. World Energy Council, September 2011, 58–59. http://www.worldenergy.org/documents/wecwei2011.pdf.

Argentine Wind Energy Association. "PE Reta: La comitiva realizó diversas actividades en Tres Arroyos" [PE Reta: the group carried out various activities in Tres Arroyos], September 5, 2011. http://www.argentinaeolica.org.ar/portal/index.php?option=comcontent&task=view&id=2837 &Itemid=3.

Asociación Mexicana de Energía Eólica (AMDEE). "Wind Energy Projects in Mexico 2011," n.d. http://amdee.org/Proyectos/Wind%20 Projects%20in%20Mexico%202011-English.pdf (accessed May 14, 2012).

Automodular Corp. "Annual Information Form." March 9, 2012. http://www.sedar.com/homepageen.htm.

Backwell, Ben. "Mainstream and Goldwind Join for 70 MW Chile Wind Farm." *Recharge*, February 2012. http://www.rechargenews.com/energy/wind/article304206.ece.

Banco Nacional De Desenvolvimento Econômico E Social (BNDES). "BNDES Approves Credit of R $358 Million for Four Wind Farms in Ceará." May 10, 2012. http://www.bndes.gov.br/SiteBNDES/ bndes /bndesen/Institucional/Press/Noticias/2012/20120 510eolicasCE.html.

———."BNDES Portfolio in PAC Totals R $123 Billion in Financing," May 7, 2010. http://www.bndes.gov.br/SiteBNDES/bndes/bndespt/ Insti tucional/SaladeImprensa/Noticias /2010/financas/20100507pac.html.

———. "Circular NQ 195, de 28 de Julho de 2006: Normas Reguladoras do Produto BNDES Finame," July 28, 2006. http://www.bndes.gov.br/SiteBNDES/export/sites/default/bndes pt/Galerias/Arquivos/produto s/ download/Circ195.pdf.

Bennett, Colin. "Latin American Wind Takes Shape." *Renewable Energy Focus*, January 19, 2011. http://www.renewableenergyfocus.com/view/ 15239/ latin-american-wind-takes-shape/.

Biller, David. "Calderón Inaugurates Iberdrola's 80MW Ventosa Wind Park." *Business News Americas*, January 23, 2009. http://www.bnamericas.com/news/electricpower/CalderoninauguratesIberdrola%27s80MWVentosawindpark.

Biofuels Journal. "Port of Duluth Handles Wind Turbine Blade Shipment to Brazil," August 26, 2008. http://www.biofuelsjournal.com/info/articles print.html?ID=62297.

Bloomberg New Energy Finance. "Chile Renewable Power Target to Spur $10 Billion in Investment." February 2, 2012.

———. Database. http://www.newenergyfinance.com (accessed various dates).

———. "IMPSA Pursuing Sales Outside Brazil as Glut Lowers Prices." February 9, 2012.

Buddelmeyer, James. "Addressing Competitiveness in the U.S. Wind Industry to Allow for Long-term Growth." Presentation at Windpower 2012, Atlanta, GA, June 5, 2012.

Business News Americas. "In Brief: DeWind Shipping Final 5 of 10 Wind Turbines – Chile," December 31, 2008. http://www.bnamericas.com/news/electricpower/DeWindshippingfinal5of10windturbi nes.

Canadian Wind Energy Association (CANWEA). "Federal Budget Fails to Extend Support for New Wind Energy Development: Canada's Ability to Compete with the U.S. for New Investment and Jobs Reduced." News release, March 4, 2010. http://www.canwea.ca/media/release/ releasee.php?newsId=76.

———. *Powering Canada's Future*. December 2011 and March 2012. http://www.canwea.ca/pdf/Canada%20Current%20Installed%20Capacity e.pdf.

Castano, Ivan. "Bolivia Plans Wind Power, Other Renewable Energy Build-out." *RenewableEnergyWorld.com*, January 5, 2011. http://www.renewableenergyworld.com/rea/news/article/2011/01/bolivia-plans-wind-powerother-renewable-energy-build-out.

Center for Clean Air Policy. "Mexico's Renewable Energy Programs," October 11, 2011.

Composite Technology Corp. "Composite Technology's DeWind Completes Turbines for Chile." News release, March 23, 2011. http://www.bloomberg.com/apps/news?pid=newsarchive&sid=acKJXTYZ1IIY.

David, Andrew S. *Shifts in U.S. Wind Turbine Equipment Trade in 2010*. U.S. International Trade Commission (USITC) Executive Briefings on Trade. Washington, DC: USITC, June 2011. http://www.usitc.gov/publications/332/executive briefings/wind EBOT commission review fi nal2.pdf.

———. *Wind Turbines: Industry and Trade Summary*. USITC, ITS-09, June 2009. http://www.usitc.gov/publications/332/ITS-2.pdf.

Diário Oficial. "Convénio ICMS 187, de 10 Dezembro de 2010" [ICMS Agreement 187], December 16, 2010. http://portal.in.gov.br/.

———. "Decreto No. 7.032, de 14 Dezembro de 2009" [Decree no. 7.032], December 15, 2009. http://portal.in.gov.br/.

———. "Portaria Interministerial No. 338, de 17 Dezembro de 1997" [Interministerial decree no. 338], December 18, 1997, 30274–30275. http://portal.in.gov.br/.

Department of Justice Canada. "Income Tax Regulations, (C.R.C., c. 945), Part XI, Capital Cost Allowance." http://laws-lois.justice.gc.ca/ eng / regulations/C.R.C., c. 945/page-44.html#h-104 (accessed April 13, 2012).

European Investment Bank (EIB). "EIB Finances Wind Farm in Mexico." News release, May 5, 2010. http://www.eib.org/projects/press/2010/2010-070-eib-finances-wind-farm-in-mexico.htm.

Export-Import Bank of the United States (Ex-Im Bank). "American Windpower Firm Grows Through Exports to Mexico as Ex-Im Back Provides First Project Financing For Wind Power." News release, November 24, 2009. http://www.exim.gov/pressrelease.cfm/2B9914C5-DF76-9BA9- 5180F9EE0A80DBCE/.

———. "Ex-Im Bank Provides $159 Million in Trade Finance for Pennsylvania Manufacturer To Export Turbines to Wind Farm in Honduras." News release, August 20, 2010. http://www.exim.gov /pressrelease.cfm/9F2BD5AC-FAFC-EE31-C412F9A785BF9C1E/.

———. Summary of Minutes of the Meeting of Board of Directors, May 31, 2012. http://www.exim.gov/article.cfm/D5E0D61A-EF48-9520-20CD 7F283467A293/.

———. "Vermont Manufacturer Makes Largest U.S. Export Of Distributed Wind Turbines, Ex-Im Bank Guarantees Financing." News release, May 26, 2011. http://www.exim.gov/pressrelease.cfm/31A72C89-950C-DB0B-5EC045872CD3F282/.

Farrell, John. *Maximizing Jobs from Clean Energy: Ontario's "Buy Local" Energy Policy*. The New Rules Project, January 2011. http://www.ilsr.org/ american-states-can-learn-much-ontarios-buy-localclean-energy-strategy/.

Fiuza, Lauro Jr. "Updates on Brazilian Markets." Presentation at Windpower 2012, Atlanta, GA, June 5, 2012.

Gamesa. "Gamesa Announces Industrial Project in Bahia and Plans to Make Brazil One of its Core Growth Markets in the Medium Term." News release, December 1, 2010. http://www.gamesacorp.com/en/ communication/news/gamesa-announces-industrial-project-in-bahia-and-plans-to-make-brazil-one-of-its-core-growth-markets-in-the-medium-term.html?idCategoria=0&fechaDesde=&especifica=0&texto=&fechaHasta.

———. "Gamesa Debuts in a New Market." News release, May 3, 2012. http://www.gamesacorp.com/en/communication/news/gamesa-debuts-in-a-new-market-uruguay-with-a-deal-to-supply-50-mw-of-turbines-to-abengoa.html?idCategoria=0&fechaDesde=&especifica=0&texto=&fecha Hasta.

———. "Gamesa Strengthens its Presence in Latin America with its First Contract in Nicaragua: To Install a Turnkey Wind Farm with Capacity of 44 MW." News release, May 28, 2012. http://www.gamesacorp.com /en/communication/news/gamesa-strengthens-its-presence-in-latin-america-with-its-first-contract-in-nicaragua-to-install-a-turnkey-wind-farm-with-capacity-of-44-mw.html?idCategoria=0&fechaDesde= &especifica=0&texto=&fechaHasta.

Gerner, Franz and Megan Hansen. *Caribbean Regional Electricity Supply Options: Toward Greater Security, Renewables and Resilience*. World Bank, 2011. https://openknowledge.worldbank.org/handle/10986/2738.

Global Trade Information Service Inc. (GTIS). Global Trade Atlas database (accessed various dates).

Global Wind Energy Council. *Global Wind Report: Annual Market Update 2011*. March 2012. http://www.gwec.net/fileadmin/documents/ News Documents/Annualreport2011lowres.pdf.

———. *Global Wind Report: Annual Market Update 2010*. April 2011. http://www.gwec.net/fileadmin/images/Publications/GWECannualmarket update2010- 2ndeditionApril2011.pdf .

———. *Global Wind Statistics 2011*. February 7, 2012. http://www.gwec.net/ fileadmin/images/News/Press/GWECGlobal Wind Statistics 2011.pdf.

High-Performance Composites. "GE Energy Announces Two New U.S. Wind Turbine Blade Facilities." *High-Performance Composites*, January 1, 2008. http://www.compositesworld.com/news/geenergy-announces-two-new-us-wind-turbine-blade-facilities.

Hydro-Québec Distribution. "Call for Tenders A/O 2009-02, Consolidated Document Integrating Addendum 1, 2, 3, 4, 5 and 6 Issued on April 30, 2010." April 30, 2010. http://www.hydroquebec.com/ distribution/en/ marchequebecois/ao-200902/pdf/consolidateddocument.pdf.

———. *Wind Power Ensuring Québec's Electricity Supply.* 3rd Quarter 2008. http://www.hydroquebec.com/publications/en/others/pdf/deplianteolienn edistribution.pdf.

Iberdrola. "Iberdrola Renovables Buys Bii Nee Stipa Wind Farm (Mexico) From Gamesa," News release, January 11, 2011. http://www.iberdrola.es /webibd/corporativa/iberdrola?IDPAG=ENMODPRENNAC11&URLPAG =/ gc/prod/en/comunicacion/notasprensa/110111NP01BiiNeeStipa.html.

IMPSA. Corporate Presentation. February 2012. http://www.impsa.com /es/ipo/Corporate%20Presentation/Corporate%20Presentation.pdf.

———. "Uruguayan UTE Approves Contract for a 50 MW Wind Project with an IMPSA Subsidiary." News release, n.d. http://www.impsa.com/en/ news/SitePages/10-04-12.aspx (accessed May 14, 2012).

International Energy Agency (IEA). *Wind: 2010 Annual Report.* 2011. http://www.ieawind.org/indexpagepostings/IEA%20Wind%202010%2 0ARcover.pdf.

Kalinoski, Gail. "Mexico's Wind Market Comes of Age." *North American Windpower*, March 2011, 42–44.

Kaplan, Matthew. "United States Wind Market Approaches Policy-Induced Cliff." North America Wind Energy Advisory, November 2, 2011. http://www.emerging-energy.com/Content/Document-Details/Wind/United-States-Wind-Market-Approaches-PolicyInduced-Cliff/1151.aspx.

Kessler, Richard A. "Siemens Sees Americas Turbine Sales." *Recharge*, June 6, 2012. http://www.rechargenews.com/energy/wind/article314826.ece.

LM Wind Power. "Announcement of Financial Results for 2010." Financial announcement, April 26, 2011. http://www.lmwindpower.com/ News/ Archive/View%20News.aspx?id=%7B3B65B3C3-C489-4738-BE37-103 9824 C3ADD%7D&y=2011.

———. "LM Wind Power Announces Site Search for Expansion in Brazil." News release, October 5, 2010. http://www.lmwindpower.com/ News/ Archive/View%20News.aspx?id=%7B9AE14519- D567-4A10-8AC5-CA71CD2B75AF%7D&y=2010.

Macquarie Group. "Macquarie Mexican Infrastructure Fund Announces Completion of Financing for 396 MW Wind Energy Project in Oaxaca, Mexico." News release, February 24, 2012. http://www.macquarie.com/ mgl/com/us/about/news/2012/20120224.

McGovern, Michael. "Mexican Orders Boost Acciona's Blade Production Start-up." *Windpower Monthly*, July 1, 2010. http://www.windpower

monthly.com/news/1013724/Mexican-orders-boostAccionas-blade-production-start-up/.

McKenna, John. "Broader Global Role Replaces Rapid Growth." *Windpower Monthly*, May 1, 2012. http://www.windpowermonthly.com/news/1129074/Broader-global-role-replaces-rapidgrowth/.

MercoPress. "Uruguay and Brazil Agree to Jointly Develop and Expand Wind Power," April 21, 2012. http://en.mercopress.com/2012/04/21/uruguay-and-brazil-agree-to-jointly-develop-andexpand-wind-power.

Ministério do Desenvolvimento, Indústria e Comércio Exterior. "Camex inclui três itens na Lista de Exceções à Tarifa Externa Comum" [Camex includes three items on the List of Exceptions to the Common External Tariff], June 18, 2009. http://www.mdic.gov.br/portalmdic/sitio/interna/noticia.php?area=1&noticia=9129.

———. "Resolução No. 07, de 1 de Março de 2007" [Resolution no. 7], March 1, 2007. http://www.mdic.gov.br//arquivo/legislacao/rescamex/2007/rescamex007-07.pdf.

Ministério de Minas e Energia. *PROINFA—EÓLICA: Guia de Habilitação Eólica* [Guide to qualifying wind energy projects], n.d. http://www.eletrobras.com/elb/data/Pages/LUMISABB61D26PTBRIE.htm (accessed April 1, 2011). "Portaria No. 242, de 25 de Junho de 2009" [Decree no. 242], June 25, 2009. http://www.mme.gov.br/mme/galerias/arquivos/legislacao/portaria/Port 242 2009.pdf.

Monteith, Laura, Ash Gupta, and Thomas J. Timmins. Gowling Lafleur Henderson LLP. "The Wind North of the 49th Parallel: U.S. Investment Opportunities through Canadian Wind Energy Tax Incentives." American Bar Association, *Energy Committees Newsletter* 8, no. 2, March 2011, 15– 17. http://www.americanbar.org/content/dam/aba/publications/nr newsletters/energy/201103 e nergy.authcheckdam.pdf.

Municipal District of Pincher Creek No. 9. Public Meeting Notice. 2012. http://www.mdpinchercreek.ab.ca.

Navigant Consulting. *Impact of the Production Tax Credit on the U.S. Wind Market*. Prepared for the American Wind Energy Association, December 12, 2011. http://www.awea.org/learnabout/publications/reports/upload/AWEA-PTC-study-121211- 2pm.pdf.

Navigant's BTM Consult. *World Market Update 2011: Forecast 2012–2016*. March 2012.

Nielsen, Stephan. "Alstom Brazil Plant to Export Wind Turbines Across Latin America." *Bloomberg*, December 8, 2011. http://www.bloomberg.

com/news/2011-12-08/alstom-brazil-plant-toexport-wind-turbines-across-latin-america.html. "Argentina Plans Biggest Wind Project with Loan from China." *Bloomberg*, July 5, 2012. http://www.bloomberg.com/news/2012-07-05/argentina-plans-biggest-wind-project-with-loan-from-china.html.

Nordic Windpower. "Nordic Windpower's First Utility Scale Wind Turbines Headed for Uruguay." News release, December 7, 2009. http://www.nordicwindpower.com/pressrelease/firstshipments.pdf.

Norfleet, Gregory R. "IWEA Head Withdraws 'Layoff' Quote." *West Branch Times*, March 2, 2012. http://www.westbranchtimes.com/article.php?viewID=7925.

North American Windpower. "DeWind to Supply Turbines to Research Site," May 26, 2011. http://www.nawindpower.com/e107 plugins/content/content.php? content.7956.

———. "First Suzlon S97 Turbines Arrive In Nova Scotia," September 29, 2011. http://www.nawindpower.com/e107plugins/ content/content.php?content.8658.

———. "Gamesa, Eolia Renovables De Inversiones to Develop Wind Farms In Mexico," December 2, 2010. http://www.nawindpower.com/e107 plugins/content/content.php?content.6965.

———. "Goldwind USA Hires Rao-Aourpally for Canada Office," June 7, 2011. http://www.nawindpower.com/e107 plugins/content/content.php?content.8018.

———. "Iberdrola Commissions 102 MW Wind Farm in Mexico," January 17, 2012. http://www.nawindpower.com/e107plugins/content/content.php?content.9228.

———. "Pattern Energy Developing 115 MW Wind Farm In Chile," November 28, 2011. http://www.nawindpower.com/e107plugins/content/content.php?content.8957.

———. "Report Reveals Top 10 Wind Turbine Manufacturers for 2011," March 26, 2012. http://www.nawindpower.com/e107plugins/content/content.php?content.9581/

———. "Vestas Nacelles Signs Deal to Meet Ontario's Local-Content Requirements," January 6, 2012. http://www.nawindpower.com/e107 plugins/content/content.php?content.9170.

———. "Vestas Receives Order for Wind Project in Mexico," July 5, 2011. http://www.nawindpower.com/naw/e107plugins/content/content.php?content.8183.

———. "Vestas to Supply Wind Turbines for 396 MW Project in Mexico," March 12, 2012. http://www.nawindpower.com/e107 plugins/content/content.php? content.9517.

Northern Colorado Business Report. "Vestas Colo. Shipping to Latin America Markets." Northern Colorado Business Report, December 16, 2011. http://www.ncbr.com/article/20111216/EDITION07/111219947.

Northern Power Systems. "Northern Power Systems Launches Canadian Wind Business." News release, September 9, 2009. http://www.northernpower. com/news/PressRelease9.9.09.php.

———. "Northern Power Systems' Wind Turbine Fleet Exceeds One Million Hours of Operation. News release, April 4, 2012. http://www.northernpower. com/news/documents/Northern-Power-Systems-Wind-Turbines-Exceeds-1-Million-Run-Hours.pdf.

Ontario Power Authority. "Feed-in Tariff Contract." Version 1.5.1, July 15, 2011. http://fit.powerauthority.on.ca/sites/default/files/FIT%20Contract%20 Version%201.5.1%20Pr ogram%20Review0.pdf,

———. Feed-in Tariff Contract Exhibits. http://fit.powerauthority.on.ca/ Storage/99/10879 FIT Contract Version 1 3 Exhibits.pdf (accessed May 4, 2010).

———. "Feed-in Tariff Program: FIT Rules." Version 1.5.1, July 15, 2011.http://fit.powerauthority.on.ca/sites/default/files/FIT%20Rules%20V ersion%201%205%201 Pro gram%20Review.pdf.

———. *Feed-In Tariff Program: Program Overview.* Version 1.1, September 30, 2009. http://fit.powerauthority.on.ca/Storage/97/10759FIT-Program-Overviewv1. 1.pdf.

———. *Ontario's Feed-in Tariff Program—Backgrounder.* April 8, 2010. http://fit.powerauthority.on.ca/Storage/100/10985Apr8BackgrounderFINAL.p df.

Patton, Dominique. "China's Wind Manufacturers Plant Seeds of Global Growth." *Recharge*, October 14, 2011. http://www.rechargenews. com/energy

Pearson, Natalie Obiko. "Suzlon Shifts Nacelle Production from Germany's Repower to Southern India." *Bloomberg*, May 20, 2011. http://www.bloomberg.com/news/2011-05-20/suzlon-shifts-nacelle-production-from-germany-s-repower-to-southern-india.html.

Phillips, Dom. "Cheap as Chips: Brazilian Wind." *Energy Economist*, no. 368 (June 2012): 20–23. http://www.gwec.net/fileadmin/images/News/ Press/EnergyEconomistmagazineinterviewRamon.pdf.

Platzer, Michaela D. *U.S. Wind Turbine Manufacturing: Federal Support for an Emerging Industry.* Congressional Research Service, R42023, September 23, 2011. http://www.fas.org/sgp/crs/misc/R42023.pdf.

Quilter, James. "Gamesa Wins First Canadian Order." *Windpower Monthly*, February 20, 2012. http://www.windpowermonthly.com/news/1118076/Gamesa-wins-first-Canadian-order.

———. "Mexico Hands Acciona $600 Million Deal for Three Wind Projects." *Windpower Monthly*, March 9, 2010. http://www.windpowermonthly.com/news/988963/Mexico-hands-Acciona600million-deal-three-wind-projects/.

RenewableEnergyWorld.com. "Gamesa Reaches 10-MW Mexican Wind Supply Deal." July 8, 2009. http://www.renewableenergyworld.com/rea/news/article/2009/07/gamesa-in-10-mwmexican-wind-supply-deal.

REpower Systems. "REpower Turbines." n.d. http://www.repower.de/fileadmin/download/unternehmen/REpowerSystemsIncCanadaactivities.pdf (accessed May 17, 2012).

reNews Americas. "Approaching Escape Velocity." *Latin America Focus*, December 16, 2010. http://renewsamericas.com/images/pdfs/reNewsAmericasLatinAmerica2010.pdf.

REVE. "La Rioja Signs Contract with IMPSA to Expand the Arauco Wind Farm." November 2, 2010. http://www.evwind.es/noticias.php?idnot=8148.

Richards, Timothy J. General Electric Co. Written statement to the House Committee on Energy and Commerce, Subcommittee on Commerce, Trade & Consumer Protection. Growing U.S. Trade in Green Technology. Hearing, October 7, 2009. http://democrats.energycommerce.house.gov/Press 111/20091007/richards testimony.pdf.

Rodriguez, Leopoldo. "Mexico Wind Industry Update." Presentation at Windpower 2012, Atlanta, GA, June 5, 2012.

———."Wind Projects in Mexico and Financing." Presentation at Windpower 2011, Anaheim, CA, May 25, 2011.

Romano, Benjamin. "GE and Vestas Slam Canadian Provinces Over Local Content." *Recharge*, October 20, 2011. http://www.rechargenews.com/energy/wind/article284605.ece.

——· "GE Boss Sees the Positives in US Wind." *Recharge*, June 1, 2012, 24–25. "Local Rules Decisive in Siemens' Plant Pick." *Recharge*, December 10, 2010. http://www.rechargenews.com/hardcopy/article239010.ece.

Sawyer, Steve. "Latin America–A Sleeping Giant Begins to Stir." *Windpower Monthly*, August 3, 2010. http://www.windpowermonthly. Com /news /1020130/Latin-America---sleeping-giant-beginsstir/.

Sciaudone, Christiana. "Argentina's Impsa Close to Deal on Uruguay Wind Farm Trio." *Recharge*, February 7, 2012. http://www.rechargenews.com/ energy/wind/article301244.ece."BNDES Audit Leaves 2 GW of Brazilian Wind Projects on Hold." *Recharge*, July 19, 2012. http://www. rechargenews.com/business area/finance/article 318333.ece. "IMPSA to Build Concrete Tower Plant in Uruguay." *Recharge*, June 1, 2012, 11 "Suzlon Puts $3bn Wind Proposal to Mexicans." *Recharge*, June 1, 2012, 5.

"Uruguay Wind Power Tender Results Delayed to January." *Recharge*, December 14, 2010. http://www.rechargenews.com/energy/ wind/article 301244.ece.

"UTE Sticking to its Guns over Tender Contracts." *Recharge*, March 18, 2011. http://www.rechargenews.com/hardcopy/article249094.ece.

Shannon, Greg. "Canadian Renewable & Conservation Expense ('CRCE') 'Green' Energy Tax Incentives." Miller Thomson LLP, January 2006. http://canambar.michaelvacirca.com/wpcontent/uploads/2011/05/AB GPS Canadian Renewable Conservation Expense 20January 2 02.pdf.

Siemens. "Hutchinson Nacelle Assembly Facility." 2011. http://www. kansascommerce.com/DocumentView.aspx?DID=814.

———."Siemens selects Tillsonburg, Ontario, as New Home for Canadian Wind Turbine Blade Manufacturing Facility." News release, December 2, 2010. http://www.siemens.ca/web/

Simpson, Scott. "Wind Power Sector Eager to Deliver on B.C. Power Promise." *Vancouver Sun*, February 3, 2012. http://www.vancouversun. com/story print.html?id=6100551&sponsor.

Sinovel. "Desenvix Taps Chinese Bank for $55m to Fund Brazil Wind Farm." News release, February 1, 2012. http://www.sinovel. com/en/ newscontent.aspx?ID=382.

———. "Sinovel Led Chinese Companies for Debut on American International Wind Energy Exhibition." News release, May 25, 2011. http://www.sinovel.com/en/newscontent.aspx?ID=270.

Soto, Onell R. "Del Mar Firm Signs Deal for Mexican Wind Farm." *San Diego Union-Tribune*, May 20, 2010. http://www.utsandiego.com/news/ 2010/may/20/del-mar-firm-signs-deal-for-mexicanwind-farm.

Suzlon. "First Suzlon S97 Turbines Arrive in North America for Installation." News release, September 28, 2011. http://webfarm.bloomberg.com/apps/news?pid=conewsstory&tkr=6YS:GR&sid=alf7mXv3eF60.

Stephens, Ken and John Green. "Hutch Can Soar with Siemens Plant." *Hutchinson News*, December 4, 2010. http://www.hutchnews.com/Print/kensiemensstory120610.

Unison. "Unison Co., Ltd. Celebrating Ceremony for the 1st Wind Farm in Jamaica." News release, October 18, 2010. http://www.unison.co.kr/Eng/Cyber/Unisnews/BoardView.aspx?idx=110&page=1&keyfield=&keyword=&TBLNM=UNISONNOTBRDENG&Category.

U.S. Department of Commerce (USDOC), International Trade Administration (ITA). *Doing Business in Brazil: 2011 Country Commercial Guide for U.S. Companies*. n.d. http://export.gov/brazil/static/CCBRCCG Full DocumentLatestegbr034878.pdf (accessed July 20, 2012).

U.S. Government Accountability Office (GAO). *Export-Import Bank: Reaching New Targets for Environmentally Beneficial Exports Presents Major Challenges for Bank*. GAO-10-682, July 2010. http://www.gao.gov/new.items/d10682.pdf.

U.S. International Trade Commission (USITC). Interactive Tariff and Trade DataWeb (DataWeb). http://dataweb.usitc.gov (accessed various dates).

Vestas. *Annual Report 2011*. 2012. http://www.vestas.com/en/annual-report-2011.aspx.

———.*Track Record as of 31 December 2011*. 2012. http://www.vestas.com/en/investor/financialreports/financial-reports-2011.aspx.

———. "Vestas Bolsters Colorado Economy: 1,000 Jobs Created, Hiring More Local Workers." News release, July 13, 2010. http://www.vestas.com/files//Filer/EN/Pressreleases/Local/ 2010/AM100713LPRUK05.pdf.

———. "Vestas Receives 99 MW Order in British Columbia, Canada." News release, December 21, 2010. http://www.vestas.com/files//Filer/EN/Pressreleases/Local/2010/AM101221NRUK.pdf.

———. "Vestas Receives 104 MW Order in Canada." News release, April 20, 2011. http://www.vestas.com/en/media/news/news-display.aspx?action=3&NewsID=2646.

Voorhis, Dan. "Energy Industry Fears U.S. Tax Credit Won't be Renewed." *The Columbus Dispatch*, April 5, 2012. http://www.dispatch.com/content/stories/business/2012/04/05/no-wind-break.html.

Waggoner, Tom. "Achieving Supply Chain Success in an Unpredictable Wind Industry." Presentation at Windpower 2012, Atlanta, GA, June 5, 2012.

Weiss, Pedro. “The Challenge and Potential for Renewable Energy in Peru.” Presentation, September 2011. http://www.iamericas.org/ presentations/ energy/ 2011-LimaRT/Pedro Weiss.pdf.

Wilkinson, Abigail. “Future Uncertain for Genren II Renewables Round.” Business News Americas, April 12, 2012. http://www.bnamericas. com/news/electricpower/future-uncertain-for-genren-iirenewables-round.

———. “Government Mulls Wind Equipment Manufacturing Measures-Impsa.” Business News Americas, March 7, 2012. http://www. bnamericas.com/news/ electricpower/government-mullswind-equipment-manufacturing-measures-impsa.

Windpower Intelligence. “Cuba: Gibara II Wind Farm Goes Online,” October 5, 2010. http://www.windpowerintelligence.com/article/HWNVFcBkNw/ 2010/10/05/cuba gibara ii win d farm goes online/.

———. “Last Clipper Windpower Turbine Installed at 80 MW Oaxaca Complex,” March 9, 2012. http://www.windpowerintelligence.com/ article/nzQyypp0cA/2012/03/09/mexico last clipper windpower turbine installed at 80mw oaxa.

———. “Mexico: Acciona Commissions Latin America's Largest Wind Complex,” March 8, 2012. http://www.windpowerintelligence.com/ article/RJBWMQmytb/2012/03/08/mexico acciona co mmissions latin americas largest wind compl/.

Windpower Monthly. “Create Opportunities When the Market Dips.” Special Report, May 2012, 21–25.

———. “Interview with Brazilian Wind Industry Vice President Lauro Fiuza,” August 13, 2010. http://www.windpowermonthly.com/news/1022252/ Interview-Brazilian-wind-industry-vicepresident-Lauro-Fiuza/.

———. “Vestas Breaks New Ground in Mexico,” March 15, 2012.
http://www.windpowermonthly.com/go/northamerica/news/1122534/Vestas-breaks-newground-Mexico

Wiser, Ryan and Mark Bolinger. *2010 Wind Technologies Market Report.* U.S. Department of Energy, Office of Energy Efficiency and Renewable Energy, June 2011. http://www1.eere.energy

Wong, Cynthia and Anthony Zlahtic. “Insight into the Current Status of the Supply Chain.” Presentation at the Ontario Feed-in Tariff Supply Chain Forum, Allstream Centre, Toronto, October 5, 2010. http://borealsolar. com/pdfs/FIT/Capital%20Power%20-%20Anthony%20Zhlatic.pdf.

World Trade Organization (WTO). “Dispute Settlement: Dispute DS412, Canada—Certain Measures Affecting the Renewable Energy Generation

Sector." September 13, 2010. http://www.wto.org/english/tratope/ dispue/ casese/ds412e.htm (accessed July 19, 2012).

———. "Dispute Settlement: Dispute DS426, Canada—Measures Relating to the Feed-in Tariff Program." August 11, 2011. http://www.wto.org/ english/tratope/dispue/casese/ds426e.htm (accessed July 19, 2012).

———. Tariff Download Facility. http://tariffdata.wto.org (accessed various dates).

World Wind Energy Association (WWEA). *World Wind Energy Report 2010*. April 2011. http://www.wwindea.org/home/ images/stories/ pdfs/world windenergyreport2010s.pdf.

———. *World Wind Energy Report 2009*. March 2010. http://www.wwindea. org/home/images/stories/worldwindenergyreport2009s.pdf.

Zindler, Ethan. Bloomberg New Energy Finance. Written statement to the Senate Committee on Finance, Subcommittee on Energy, Natural Resources, and Infrastructure. Renewable Energy Tax Incentives: How have the Recent and Pending Expirations of Key Incentives Affected the Renewable Energy Industry in the United States? Hearing, April 27, 2012. http://finance.senate.gov/imo/media-03-27senatefinancetestimonyzindler.p df.

End Notes

[1] This paper represents solely the views of the authors and is not meant to represent the views of the U.S. International Trade Commission or any of its commissioners. The invaluable assistance of Michael Anderson, Monica Reed, Peg Hausman, Karen Laney, and Wanda Tolson is gratefully acknowledged.

[2] For the purposes of this paper, utility-scale wind turbines are those rated at 100 kilowatts (kW) or more. While 100 kW turbines are on the line between small and large wind turbines, they are included here as there are significant U.S. exports of 100 kW wind turbines. The paper will also focus on on-shore wind turbines, as there is not currently any U.S. manufacturing of off-shore wind turbines.

[3] Wind-powered generating sets include nacelles and any items imported with the nacelle, such as the blades or hub. If these components are imported or exported separately from the nacelle, they are included in different HS subheadings. Wind-powered generating sets are included in 8502.31 in the international Harmonized Commodity Description and Coding System (HS).

[4] This paper only examines exports of equipment and does not discuss service exports.

[5] Wind turbine OEMs typically design the wind turbine or license a wind turbine design, produce the nacelle, and sell the wind turbine under their own name (e.g., Gamesa, General Electric Co. (GE), Siemens, and Vestas).

[6] Some OEMs also engage in project development.

[7] The drivers of demand in Canada and Latin America will be discussed in the individual country sections. Though the demand drivers vary by country, they may include government policies (at the national or subnational level), excellent wind resources, improvements in the grid infrastructure, better availability of financing, declining wind turbine prices, rising electricity demand, the complementarity of wind and hydroelectric power, a need to reduce dependence on imported fossil fuels, and a desire to install technologies with a minimal environmental impact. Navigant's BTM Consult, World Market Update 2011, March 2012, 70; GWEC, Global Wind Report: Annual Market Update 2011, March 2012, 11; McKenna, "Broader Global Role," May 1, 2012.

[8] Navigant's BTM Consult, World Market Update 2011, March 2012, 70; GWEC, Global Wind Report: Annual Market Update 2011, March 2012, 11; McKenna, "Broader Global Role," May 1, 2012.

[9] Though China and the EU were the two largest markets in 2011, they are primarily supplied through local production, with EU imports of wind-powered generating sets from all sources totaling only \$67 million and Chinese imports totaling only \$12 million in 2011. U.S. exports to the EU were less than \$1 million in 2011 and there were no U.S. exports to China that year. The largest U.S.-based wind turbine producer, GE, has production plants in both China and Europe and many of the leading producers in the United States are EU-based firms that also have production both in their home region and China. GWEC, Global Wind Statistics 2011, February 7, 2012; GTIS, Global Trade Atlas database (accessed April 18, 2012); USITC DataWeb/USDOC (accessed April 13, 2012); David, Wind Turbines, June 2009, 20; Siemens, "Hutchinson Nacelle Assembly Facility," 2011, 7; Vestas Web site, http://www.vestas.com/en/about-vestas/company-structure.aspx (accessed December 22, 2011).

[10] In some cases, project developers may have indicated that they plan to use turbines from a particular company, but it is not clear if the developers have placed firm orders for these turbines.

[11] This discussion focuses on the PTC since it has been in place for the longest period of time and has traditionally been a key driver of installations. If the PTC is renewed, installation levels in 2013 will likely depend on the timing of its renewal. In 2014, installations with the PTC are projected to be between 6 and 8 GW. One estimate concluded that the state RPSs will drive 4–6 GW in installations during 2011–20, but a substantial portion of these could come in the second half of the decade if the PTC expires. Kaplan, "United States Wind Market," November 2, 2011, 6–7; Navigant's BTM Consult, World Market Update 2011, March 2012, 70; Zindler, written statement to the Senate Committee on Finance, March 27, 2012, 2; Voorhis, "Energy Industry Fears," April 5, 2012; AWEA, U.S. Wind Industry Fourth Quarter 2011 Market Report, January 2012, 1; AWEA, U.S. Wind Industry Year-End 2010 Market Report, January 2011, 1; AWEA, Annual Wind Industry Report: Year Ending 2008, 2009, 4; Wiser and Bolinger, 2010 Wind Technologies Market Report, June 2011, 61–62

[12] See table 1 below for more on U.S. nacelle production capacity. In some cases, production capacity is announced capacity and companies probably have not reached announced production levels. It is unclear whether U.S. demand at current levels is sufficient to sustain all of the nacelle manufacturers currently active in the U.S. market, but at current demand levels a dramatic contraction in production capacity is less likely. For discussion of the impact of the PTC expiration on U.S. industry, see, for example, Navigant Consulting, Impact of the Production Tax Credit, December 12, 2011; Zindler, written statement to the Senate Committee on Finance, March 27, 2012.

[13] Kessler, "Siemens Sees Americas Turbine Sales," June 6, 2012; Windpower Monthly, "Create Opportunities When the Market Dips," May 2012, 25; Romano, "GE Boss," June 1, 2012, 24–25; Norfleet, "IWEA Head Withdraws 'Layoff' Quote," March 2, 2012; industry representative, interview by USITC staff, June 6, 2012.

[14] Companies also sought to reduce transportation costs, minimize the risks of currency fluctuations, ease logistical challenges, and avoid import duties. David, Wind Turbines, June 2009, 6.

[15] REpower is based in Germany and is a subsidiary of India-based Suzlon. Repower was previously an independent firm and will be discussed separately from its parent company here as this will allow a more nuanced discussion of market trends.

[16] Information is not available on whether Molded Fiberglass and TPI Composites supply OEMs other than GE. LM Wind Power, "Announcement of Financial," April 26, 2011; High-Performance Composites, "GE Energy Announces," January 1, 2008

[17] The 10-digit provision in the export classification system (Schedule B) that includes blades also includes non-wind products, so blade export data will only be discussed in the specific country sections below and only if the data provide some useful insights into U.S. trade.

[18] The leading markets have fluctuated annually, with exports for a small number of projects generally dictating which markets accounted for the most exports. Though the EU is not discussed extensively in this paper, Northern Power Systems has had success in exporting its 100 kW turbine to Italy, where there are incentives for the installation of wind power projects of less than 200 kW. USITC DataWeb/USDOC (accessed February 14, 2012); Ex-IM Bank, "Vermont Manufacturer," May 26, 2011.

[19] GTIS, Global Trade Atlas database (accessed March 29, 2012).

[20] Northern Colorado Business Report, "Vestas Colo. Shipping to Latin America Markets," December 16, 2011.

[21] Percentage is of the customs value. USITC DataWeb/USDOC (accessed April 27, 2012).

[22] Consolidated Appropriations Act, 2012, Pub L. No. 112-74, § 5, 125 Stat. 1191 (December 23, 2011).

[23] GAO, Export-Import Bank, July 2010, 4, 39–40; industry representative, interview by USITC staff, April 24, 2012; Richards, written statement to the House Committee on Energy and Commerce, October 7, 2009, 10.

[24] Local content requirements and the specific impact of the requirements will be discussed in more detail in the individual country sections.

[25] Tariffs will be discussed in more detail in the individual country sections.

[26] CANWEA, Powering Canada's Future, December 2011.

[27] Navigant's BTM Consult, World Market Update 2011, March 2012, 70.

[28] CANWEA, Powering Canada's Future, December 2011.

[29] See discussion of project pipelines below.

[30] Simpson, "Wind Power Sector Eager to Deliver," February 3, 2012.

[31] BC Hydro Web site, http://www.bchydro.com/energyinbc/acquiringpower/closedofferings/cleanpowercall/outcome.html (accessed April 10, 2012); Simpson, "Wind Power Sector Eager to Deliver," February 3, 2012.

[32] Natural Resources Canada, "Overview of Canada's Energy Policy," n.d., http://www.nrcan.gc.ca/energy/policy/1352 (accessed May 1, 2012).

[33] Alberta is the only province without a provincial-owned utility.

[34] Ontario Power Authority, Ontario's Feed-in Tariff Program—Backgrounder, April 8, 2010.

[35] Ontario Power Authority Web site, http://www.powerauthority.on.ca/current-electricity-contracts/wind-power (accessed April 13, 2012).

[36] Farrell, Maximizing Jobs from Clean Energy, January 2011, 12; Romano, "GE and Vestas Slam Canadian Provinces," October 20, 2011.

[37] Hydro-Québec, Wind Power Ensuring Québec's Electricity Supply, 3rd Quarter 2008, 5; Hydro-Québec, "Call for Tenders A/O 2009-02," April 30, 2010, 7; Hydro-Québec Web site, http://www.hydroquebec.com (accessed April 6, 2012).

[38] Ontario Power Authority, Feed-In Tariff Program, Program Overview, Version 1.1, September 30, 2009, 5; Ontario Power Authority Web site, http://fit.powerauthority.on.ca/domestic-content-1 (accessed April 12, 2012).

[39] Ontario Power Authority, FIT Contract Exhibits (accessed May 4, 2010).

[40] WTO, "Dispute Settlement: Dispute DS412," September 13, 2010.

[41] The panel reports are expected to be issued in November 2012. WTO, "Dispute Settlement: Dispute DS426," August 11, 2011.

[42] Government of Canada Web site, http://www.ecoaction.gc.ca/ecoenergy-ecoenergie/power-electricite/index-eng.cfm (accessed April 12, 2012); CANWEA, "Federal Budget Fails to Extend Support," March 4, 2010.

[43] Department of Justice Canada, Income Tax Regulations, CRC, c. 945 (2012), part XI, "Capital Cost Allowance," http://lawslois.justice.gc.ca/eng/regulations/C.R.C.,c.945/page-44.html #h-104 (accessed April 13, 2012); Monteith, Gupta, and Timmins, "The Wind North Of The 49th Parallel," March 2011, 15–17.

[44] Shannon, "Canadian Renewable & Conservation Expense ('CRCE')," January 2006; Canada Revenue Web site, http://www.cra-arc.gc.ca/tx/bsnss/tpcs/fts-paa/nrgy-eng.html (accessed April 13, 2012).

[45] Upfront developmental costs, such as engineering and design work and feasibility studies could be deducted as an expense immediately or carried forward indefinitely. Under a "flow-through" shares financing system, eligible expenses may be transferred or sold to shareholders (i.e., investors). The investors would then have the full benefit of the income tax expenses (which they can use to reduce their tax liability), while the company selling the shares receives money to finance its activities.

[46] CANWEA Web site, http://www.canwea.ca/farms/wind-farmse.php (accessed September 7, 2011 and April 2, 2012).

[47] Romano, "Local Rules Decisive," December 10, 2010.

[48] The only project using Chinese wind turbines that was identified was a 60 MW wind project in Alberta that is planning to use Sinovel wind turbines. North American Windpower, "Goldwind USA Hires Rao-Aourpally," June 7, 2011; Sinovel, "Sinovel Led Chinese Companies," May 25, 2011; Municipal District of Pincher Creek No. 9, Public Meeting Notice, 2012.

[49] The only project using Indian wind turbines that was identified was a 31.5 MW wind project in Amherst that is using Suzlon wind turbines. Suzlon announced in 2011 that it was shifting production of some REpower nacelles to India; if these nacelles are used for projects in Canada rather than those produced in Europe, there could be additional Canadian imports from India. Suzlon, "First Suzlon S97 Turbines Arrive," September 28, 2011; Pearson, "Suzlon Shifts Nacelle Production," May 20, 2011.

[50] Industry representative, e-mail to USITC staff, October 1, 2010; USITC DataWeb/USDOC (accessed April 10, 2012); CANWEA Web site, http://www.canwea.ca/farms/wind-farmse.php (accessed September 7, 2011 and April 2, 2012); Northern Power Systems, "Northern Power Systems Launches," September 9, 2009.

[51] GTIS, Global Trade Atlas database (accessed April 2, 2012).

[52] David, Shifts in U.S. Wind Turbine, June 2011.

[53] Hydro-Québec, Wind Power Ensuring Québec's Electricity Supply, 3rd Quarter 2008, 5.

[54] USITC DataWeb/USDOC (accessed April 10, 2012).

[55] Industry representative, e-mail to USITC staff, October 1, 2010; USITC DataWeb/USDOC (accessed April 10, 2012); CANWEA Web site, http://www.canwea.ca/farms/wind-farmse.php (accessed September 7, 2011, and April 2, 2012); Norfleet, "IWEA Head Withdraws 'Layoff' Quote," March 2, 2012.

[56] Quilter, "Gamesa Wins First Canadian Order," February 20, 2012; AWEA, "Gamesa Turbines from Pa. Plant," February 23, 2012; North American Windpower, "DeWind to Supply Turbines to Research Site," May 26, 2011.

[57] Vestas, "Vestas Receives 104 MW Order," April 20, 2011; industry representative, interview by USITC staff, September 7, 2010; Vestas, "Vestas Bolsters Colorado Economy," July 13, 2010; Vestas, "Vestas Receives 99 MW Order," December 21, 2010; Capital Power Web site, http://www.capitalpower.com/community/consultationengagement/Pages/ qualitywind project.aspx and http://www.capitalpower.com/community/ consultationengagement/ Documents/Quality-Wind/2012-spring-quality-windinfo-sheet.pdf (accessed July 11, 2012).

[58] Stephens and Green, "Hutch Can Soar," December 4, 2010.

[59] GTIS, Global Trade Atlas database (accessed April 10, 2012).

[60] This provision, 8412.90.9080 (parts of generators n.e.s.o.i.), is even less specific to wind than Canada's import data for parts of windmills. However, comparing month and port of export with that for wind-powered generating sets, it appears that at least a portion of the exports were correlated with nacelle exports and that at least some of these products were blades, hubs, and/or other wind components.

[61] Vestas, "Vestas Receives 104 MW Order," April 20, 2011; industry representative, interview by USITC staff, September 7, 2010; Vestas, "Vestas Bolsters Colorado Economy," July 13, 2010; Vestas, "Vestas Receives 99 MW Order," December 21, 2010; Capital Power Web site, http://www.capitalpower.com/community/consultationengagement/ Pages/ qualitywind project.aspx and http://www.capitalpower.com/community/ consultationengagement/ Documents/Quality-Wind/2012-spring-quality-wind- info-sheet.pdf (accessed July 11, 2012).

[62] REpower, "REpower Turbines," n.d. (accessed May 17, 2012).

[63] Romano, "Local Rules Decisive," December 10, 2010; Siemens, "Siemens Selects Tillsonburg," December 2, 2010; Kessler, "Siemens Sees Americas Turbine Sales," June 6, 2012.

[64] GWEC, Global Wind Report: Annual Market Update 2010, April 2011, 11, 49; AMDEE, "Wind Energy Projects," n.d. (accessed May 14, 2012); GWEC, Global Wind Statistics 2011, February 7, 2012; media reports, press releases, and company Web sites.

[65] Navigant's BTM Consult, World Market Update 2011, 70.

[66] AMDEE, "Wind Energy Projects," n.d. (accessed May 14, 2012); Kalinoski, "Mexico's Wind Market," March 2011, 42

[67] GWEC, World Market Update 2010, 2011, 48; IEA, Wind: 2010 Annual Report, 2011, 124; Kalinoski, "Mexico's Wind Market," March 2011, 42–44.

[68] Rodriguez, "Wind Projects in Mexico and Financing," May 25, 2011; GWEC, World Market Update 2010, 48; Center for Clean Air Policy, "Mexico's Renewable Energy Program," October 11, 2011, 3; Bierzwinsky, "Renewables Opportunities in Mexico," June 2011, 16–17.

[69] IEA, Wind: 2010 Annual Report, 2011, 126; Rodriguez, "Wind Projects in Mexico and Financing," May 25, 2011; Kalinoski, "Mexico's Wind Market," March 2011, 44; Center for Clean Air Policy, "Mexico's Renewable Energy Program," October 11, 2011, 7–10.

[70] GWEC, World Market Update 2010, 2011, 48–49; IEA, Wind: 2010 Annual Report, 2011, 125.

[71] GWEC, Annual Wind Report 2010, 49; Rodriguez, "Wind Projects in Mexico and Financing," May 25, 2011; Kalinoski, "Mexico's Wind Market," March 2011, 41.

[72] Kalinoski, "Mexico's Wind Market Comes of Age," March 2011, 44; Center for Clean Air Policy, "Mexico's Renewable Energy Program," October 11, 2011, 13.

[73] Rodriguez, "Wind Projects in Mexico and Financing," May 25, 2011.

[74] AMDEE, "Wind Energy Projects," n.d. (accessed May 14, 2012); Kalinoski, "Mexico's Wind Market," March 2011, 44; Quilter, "Mexico Hands Acciona," March 9, 2010; Windpower Monthly, "Vestas Breaks New Ground," March 15, 2012.

[75] Rodriguez, "Wind Projects in Mexico and Financing," May 25, 2011; Kalinoski, "Mexico's Wind Market," March 2011, 44; EIB, "EIB finances wind farm in Mexico," May 5, 2010; Ex-Im Bank, "American Windpower Firm Grows," November 24, 2009; Macquarie Group, "Macquarie Mexican Infrastructure Fund Announces," February 24, 2012; Bierzwinsky, "Renewables Opportunities in Mexico," June 2011, 18.

[76] No projects were completed in 2007–08.

[77] These firms' project pipelines are discussed in more detail below.

[78] Sciaudone, "Suzlon Puts $3bn Wind," June 1, 2012, 5; Rodriguez, "Mexico Wind Industry Update," June 5, 2012.

[79] Mitsubishi Power Systems Web site, http://www.mpshq.com/facilities/vientek.html (accessed March 21, 2012); AMDEE, "Wind Energy Projects," n.d. (accessed May 14, 2012).

[80] GTIS, Global Trade Atlas database (accessed March 21, 2012).

[81] USITC DataWeb/USDOC (accessed March 20, 2012); AMDEE, "Wind Energy Projects," n.d. (accessed May 14, 2012).

[82] GTIS, Global Trade Atlas database (accessed March 21, 2012).

[83] GTIS, Global Trade Atlas database (accessed March 20, 2012); Vestas, Annual Report 2011, 2012, 16.

[84] AMDEE, "Wind Energy Projects," n.d. (accessed May 14, 2012); Biller, "Calderón Inaugurates," January 23, 2009; North American Windpower, "Iberdrola Commissions," January 17, 2012; Iberdrola, "Iberdrola Renovables Buys," January 11, 2011; RenewableEnergyWorld.com, "Gamesa Reaches 10-MW Mexican Wind Supply Deal," July 8, 2009.

[85] The one project in Mexico funded by the Ex-Im Bank uses U.S.-made Clipper wind turbines. EIB, "EIB Finances Wind Farm," May 5, 2010; North American Windpower, "Iberdrola Commissions," January 17, 2012; Ex-Im Bank, "American Windpower Firm Grows," November 24, 2009; North American Windpower, "Vestas To Supply Wind Turbines," March 12, 2012; Macquarie Group, "Macquarie Mexican Infrastructure Fund Announces," February 24, 2012.

[86] For example, the projects in Mexico that have used Clipper wind turbines to this point were developed by Mexico-based Fuerza Eolica and France-based EDF Energies, while U.S.-based Cannon Power Group plans to source turbines for its projects from Gamesa's U.S. plant. Spanish firms have played a role in the development of a number of projects, many of which have sourced a significant share of their nacelles from the EU. AMDEE, "Wind Energy Projects," n.d. (accessed May 14, 2012); Soto, "Del Mar Firm Signs Deal," May 20, 2010.

[87] Soto, "Del Mar Firm Signs Deal," May 20, 2010; North American Windpower, "Gamesa, Eolia Renovables De Inversiones," December 2, 2010; United Nations Framework

Convention on Climate Change (UNFCCC) Clean Development Mechanism (CDM) Web site, http://cdm.unfccc.int (accessed April 30, 2012).

[88] Vestas, "Vestas Bolsters Colorado Economy," July 13, 2010; industry representative, interview by USITC staff, September 7, 2010; Soto, "Del Mar Firm Signs Deal," May 20, 2010; GTIS, Global Trade Atlas database (accessed July 19, 2012). Project pipeline from AMDEE, "Wind Energy Projects," n.d. (accessed May 14, 2012), and other projects identified by USITC staff.

[89] The estimated project pipeline and project construction data here are a conservative estimate that includes only projects listed by the Mexican Wind Energy Association with secured access to transmission capacity, together with a small number of additional projects identified by USITC staff. Soto, "Del Mar Firm Signs Deal," May 20, 2010; Kalinoski, "Mexico's Wind Market," March 2011, 42.

[90] Little information is available on U.S. blade exports to this this point. U.S. exports to Mexico in the provision that includes blades were $18.4 million in 2007, $8.1 million in 2008, $8.7 million in 2009, $109.5 million in 2010, and $27.4 million in 2011. The increase in exports in 2010 could be due to exports of blades or other wind components, but it is not possible to be certain. USITC DataWeb/USDOC (accessed April 12, 2012).

[91] Soto, "Del Mar Firm Signs Deal," May 20, 2010.

[92] Vestas, "Vestas Bolsters Colorado Economy," July 13, 2010.

[93] Acciona, "Acciona Opens," June 30, 2010; McGovern, "Mexican Orders Boost," July 1, 2010.

[94] GWEC, Global Wind Report: Annual Market Update 2011, March 2012, 24.

[95] Windpower Monthly, "Interview with Brazilian Wind Industry," August 13, 2010.

[96] Ministério de Minas e Energia Web site, http://www.mme.gov.br/programas/proinfa (accessed May 17, 2010).

[97] GWEC, Global Wind Report: Annual Market Update 2011, March 2012, 24.

[98] Wind is expected to increase from 1.3 percent of electricity generation capacity in 2011 to 12.5 percent in 2020. GWEC, Global Wind Report: Annual Market Update 2011, March 2012, 24–25; Fiuza, "Updates on Brazilian Markets," June 4, 2012.

[99] GWEC, Global Wind Report: Annual Market Update 2011, March 2012, 25.

[100] BNDES offers low cost financing and can finance up to 70 percent of a wind project. Phillips, "Cheap as Chips," June 2012, 22.

[101] BNDES, "BNDES Approves," March 3, 2011.

[102] BNDES, "BNDES Portfolio," May 7, 2010.

[103] The exemption is granted by the issued by Brazilian Committee of the Treasury Policy (Conselho Nacional de Política Fazendária, or CONFAZ). Diário Oficial, "Portaria Interministerial No. 338," December 18, 1997, 30274–30275; USDOC, ITA, Doing Business in Brazil, n.d., 52 (accessed July 20, 2012).

[104] Diário Oficial, "Convénio ICMS 187," December 16, 2010, 48.

[105] Diário Oficial, "Decreto No. 7.032," December 15, 2009, 64–65; USDOC, Doing Business in Brazil, n.d., 52 (accessed July 20, 2012).

[106] GTIS, Global Trade Atlas database (accessed May 1, 2012).

[107] Gamesa, "Gamesa Announces Industrial Project," December 1, 2010; Nielsen, "Alstom Brazil Plant," December 8, 2011; LM Wind Power, "LM Wind Power Announces Site Search," May 10, 2010.

[108] Tecsis Web site, http://www.tecsis.com.br/site/Wind.html (accessed July 22, 2012).

[109] Sinovel, "Desenvix Taps Chinese Bank," February 1, 2012.

[110] USITC DataWeb/USDOC (accessed July 20, 2012).

[111] LM also exported 54 blades from North Dakota to Brazil in 2008. Duluth Seaway Port Authority, "Exporting Blades," June 28, 2012; Ex-IM Bank, Summary of Minutes, May 31, 2012; Biofuels Journal, "Port of Duluth," August 26, 2008.

[112] Ministério de Minas e Energia, PROINFA – EÓLICA: Guia de Habilitação Eólica, n.d., 11, 36–37 (accessed April 1, 2011);

[113] BNDES, "Circular Nº 195," July 28, 2006.

[114] Ministério de Minas e Energia, Gabinete do Ministro, "Portaria No. 242," June 25, 2009.

[115] Additional firms may also be affected, but media reports vary on the other firms that were noncompliant. Sciaudone, "BNDES Audit," July 19, 2012.

[116] Ministério do Desenvolvimento, Indústria e Comércio Exterior, "Resolução N° 07," March 1, 2007.

[117] This coincided with the action by the Ministry of Mines and Energy noted above. Ministério do Desenvolvimento, Indústria e Comércio Exterior, "Camex inclui três itens na Lista de Exceções à Tarifa Externa Comum," June 18, 2009.

[118] GWEC, Global Wind Statistics 2011, February 7, 2012.

[119] Navigant's BTM Consult, World Market Update 2011, March 2012, 70.

[120] GWEC, World Market Update 2010, 2011, 28–29; Bennett, "Latin American Wind Takes Shape," January 19, 2011; Bloomberg New Energy Finance, "Chile Renewable Power Target," February 2, 2012; Sawyer, "Latin America–A Sleeping Giant," August 3, 2010; Castano, "Bolivia Plans Wind Power," January 5, 2011; Araluce, "Clean Energy," September 2011, 58; Weiss, "The Challenge and Potential for Renewable Energy," September 2011, 6–8.

[121] Castano, "Bolivia Plans Wind Power," January 5, 2011.

[122] GWEC, World Market Update 2010, 2011, 29; Bennett, "Latin American Wind Takes Shape," January 19, 2011; Bloomberg New Energy Finance, "Chile Renewable Power Target," February 2, 2012; Sawyer, "Latin America–A Sleeping Giant," August 3, 2010; reNews Americas, "Approaching Escape Velocity," December 16, 2010.

[123] Bloomberg New Energy Finance database (accessed March 22, 2010); Vestas, Track Record as of 31 December 2011, 2012, 4–5.

[124] Bloomberg New Energy Finance database (accessed March 22, 2010).

[125] Impsa also has a wind turbine plant in Brazil and wind-related production in Malaysia. IMPSA Web site, http://www.impsa.com (accessed April 17, 2012); industry representative, interview by USITC staff, Atlanta, GA, June 5, 2012.

[126] Bloomberg New Energy Finance database (March 22, 2012).

[127] Bloomberg New Energy Finance, "IMPSA Pursuing Sales," February 9, 2012.

[128] Patagonia NRG, http://www.nrgpatagonia.com/ (accessed April 17, 2012); INVAP Web site, http://www.invap.com.ar (accessed April 17, 2012).

[129] I.T.P. Web site, http://www.itp.com.pa/index.html (accessed April 17, 2012); Turbowinds Web site, http://www.turbowinds.com/index.php (accessed April 17, 2012).

[130] Backwell, "Mainstream and Goldwind," February 2012; Patton, "China's Wind Manufacturers," October 14, 2011; Windpower Intelligence, "Cuba: Gibara II," October 5, 2010.

[131] GTIS, Global Trade Atlas database (accessed April 18, 2012); Argentine Wind Energy Association, "PE Reta: La Comitiva Realizó Diversas Actividades," [PE Reta: the group carried out various activities], September 5, 2011.

[132] U.S.-based Nordic Windpower, which no longer has an operational plant, also installed three wind turbines in Uruguay. Nordic Windpower, "Nordic Windpower's First Utility Scale

Wind Turbines," December 7, 2011; GTIS, Global Trade Atlas database (accessed April 18, 2012).

[133] Unison, "Unison Co., Ltd. Celebrating Ceremony," October 18, 2010; United Nations Framework Convention on Climate Change (UNFCCC) Clean Development Mechanism (CDM) Web site, http://cdm.unfccc.int (accessed April 17, 2012).

[134] North American Windpower, "Pattern Energy Developing 115 MW," November 28, 2011; UNFCCC CDM Web site, http://cdm.unfccc.int (accessed April 17, 2012); Bloomberg New Energy Finance database (accessed March 22, 2012); REVE, "La Rioja Signs Contract with IMPSA," November 2, 2010; IMPSA, "Uruguayan UTE Approves Contract," n.d.; IMPSA, Corporate Presentation, February 2012, 14.

[135] Bloomberg New Energy Finance database (accessed March 22, 2012); GTIS, Global Trade Atlas database (accessed March 23, 2012); Ex-Im Bank, "Export-Import Bank Provides $159 Million," August 20, 2010; Composite Technology Corp., "Composite Technology's DeWind," March 23, 2011.

[136] Based on information compiled by USITC staff.

[137] WTO Tariff Download Facility (accessed March 22, 2012 and April 17, 2012); text of U.S.-Columbia FTA.

[138] Wilkinson, "Future Uncertain for Genren II Renewables Round," April 12, 2012.

[139] Bloomberg New Energy Finance database (March 22, 2012 and July 11, 2012); GTIS, Global Trade Atlas database (accessed April 18, 2012); Argentine Wind Energy Association, "PE Reta: La Comitiva Realizó Diversas Actividades en Tres Arroyos," [PE Reta: the group carried out various activities], September 5, 2011; Gamesa, "Gamesa Announces Industrial Project," December 1, 2010; Nielsen, "Alstom Brazil Plant," December 8, 2011; industry representative, interview by USITC staff, Atlanta, GA, June 5, 2012; Nielsen, "Argentina Plans," July 5, 2012; press releases, media reports, and company Web sites.

[140] Information is not available on whether Vestas intends to supply this market from the United States or another location. Based on data compiled by USITC staff and from Bloomberg New Energy Finance database (accessed March 22, 2012).

[141] While nacelles now enter Argentina duty free, the government is reportedly considering ways to promote the development of a domestic industry. According to one report, options might include "increased import duties for the equipment, a minimum requirement of locally produced goods or... an incentive program for local manufacturers." Wilkinson, "Government Mulls Wind," March 7, 2012.

[142] MercoPress, "Uruguay and Brazil Agree to Jointly Develop and Expand Wind Power," April 21, 2012.

[143] Based on data compiled by USITC staff.

[144] Uruguay also imported three 1 MW wind turbines from U.S. producer Nordic Windpower in 2010. Industry representative, telephone interview by USITC staff, April 24, 2012; Gamesa, "Gamesa Debuts in a New Market," May 3, 2012; Nordic Windpower, "Nordic Windpower's First Utility Scale Wind Turbines," December 7, 2009; GTIS, Global Trade Atlas database (accessed April 18, 2012).

[145] Gamesa, "Gamesa Announces Industrial Project," December 1, 2010; Nielsen, "Alstom Brazil Plant," December 8, 2011; Buddelmeyer, "Addressing Competitiveness," June 5, 2012.

[146] It is not clear if local content was a factor in later wind tenders. Sciaudone, "Uruguay Wind Power Tender Results," December 14, 2010; Sciaudone, "UTE Sticking to its Guns," March 18, 2011; Sciaudone, "Argentina's Impsa," February 7, 2012; Sciaudone, "IMPSA to Build," June 1, 2012, 11.

[147] Bloomberg New Energy Finance database (accessed March 22, 2012); media reports, press releases, and company Web sites.

[148] Bloomberg New Energy Finance database (accessed March 22, 2012); Composite Technology Corp., "Composite Technology's DeWind," March 23, 2011.

[149] Several producers, though none of the firms just mentioned, have indicated that they intend to supply the market in Chile from Brazil. Nielsen, "Alstom Brazil Plant," December 8, 2011; Gamesa, "Gamesa Announces Industrial Project," December 1, 2010; GTIS, Global Trade Atlas database (accessed July 11, 2012).

[150] WTO Tariff Download Facility (accessed March 22, 2012 and April 17, 2012); text of U.S.-Chile FTA.

[151] Text of the U.S.-Panama trade promotion agreement.

[152] Based on data compiled by USITC staff.

[153] Vestas has indicated that some of the components for at least one of the projects it is supplying in Nicaragua will be sourced from the United States and Gamesa is planning to produce some of the turbines for a Nicaraguan project in the United States. It is not clear exactly which components these firms plan to source from the United States. Northern Colorado Business Report, "Vestas Colo. Shipping," December 16, 2011; Gamesa, "Gamesa Strengthens its Presence," May 28, 2012.

[154] Unison, "Unison Co., Ltd. Celebrating Ceremony," October 18, 2010.

[155] Based on data compiled by USITC staff and from Bloomberg New Energy Finance database (accessed March 22, 2012).

[156] Gerner and Hansen, Caribbean Regional Electricity Supply Options, 2011, 6.

[157] Gerner and Hansen, Caribbean Regional Electricity Supply Options, 2011, 22.

[158] USITC DataWeb/USDOC (accessed April 27, 2012); Northern Power Systems, ""Northern Power Systems' Wind Turbine Fleet," April 4, 2012.

[159] Ex-Im Bank, "Export-Import Bank Provides $159 Million," August 20, 2010; USITC DataWeb/USDOC (accessed April 17, 2012).

[160] LM Wind Power has indicated that when it completes its plant in Brazil, it intends to export from Brazil to other countries in South America. While no information is available on exports by Tecsis to other markets in Latin America, this firm is a supplier to multiple OEMs and has supplied the European and North American markets. Tecsis Web site, http://www.tecsis.com.br/site/Wind.html (accessed April 18, 2012); LM Wind Power, "LM Wind Power Announces Site Search," October 5, 2010.

In: Wind Turbines
Editor: Derrick A. Foster
ISBN: 978-1-62417-569-5
© 2013 Nova Science Publishers, Inc.

Chapter 2

UNDERSTANDING TRENDS IN WIND TURBINE PRICES OVER THE PAST DECADE*

Mark Bolinger and Ryan Wiser

EXECUTIVE SUMMARY

Berkeley Lab has gathered price data on 81 U.S. wind turbine transactions totaling 23,850 MW announced from 1997 through early 2011. Figure ES-1 depicts these reported wind turbine transaction prices (along with the associated trend line), broken out by the size of the transaction (in MW). Figure ES-1 also presents average (global) turbine prices reported by Vestas for the years 2005 through 2010, as well as a range of reported pricing (among various turbine manufacturers) for transactions signed in 2010 and so far in 2011 (with 2011 prices generally lower than 2010 prices).

After hitting a low of roughly $750/kW from 2000 to 2002, average wind turbine prices *doubled* through 2008, rising to an average of roughly $1,500/kW. Wind turbine prices have since declined substantially, with price quotes for transactions executed in 2010 and to date in 2011 ranging from $900-$1,400/kW depending on the manufacturer and turbine model. For example, turbines designed for lower wind speed sites – deploying higher hub heights and larger rotor diameters for a given nameplate capacity – are priced at the higher end of this range. These quotes suggest price declines of as much

---

* This is an edited, reformatted and augmented version of a Lawrence Berkeley National Laboratory publication, LBNL-5119E, dated October 2011.

as 33% or more since late 2008, with an average decline closer to perhaps 20% for orders announced in 2010 (as opposed to in 2011, which has seen further price declines).

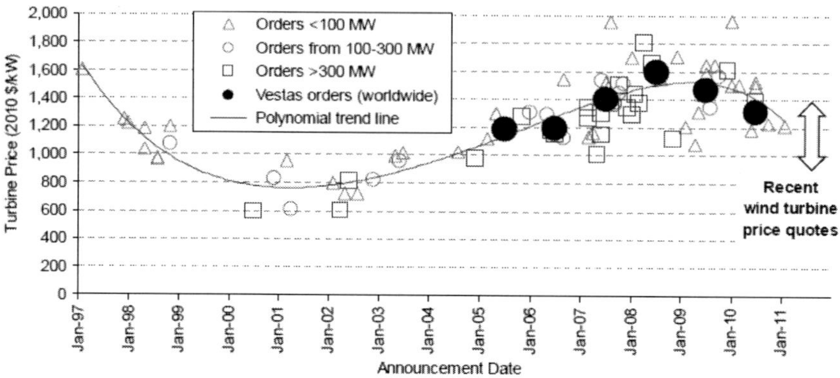

Source: Berkeley Lab, Vestas (2011b, 2011c, 2011d), Bloomberg NEF (2011b).

Figure ES-1. Wind Turbine Prices in the United States.

These two substantial and opposing wind turbine price trends over the past decade – and particularly the doubling in prices in the 2002-2008 period – run counter to the smooth, gradually declining technology cost trajectories that are often assumed by energy analysts modeling the diffusion of new technologies, including wind power. Understanding and explaining this notable discrepancy between theory and historical reality is the primary motivation for this work.

Taking a bottom-up approach, this report examines seven primary drivers of wind turbine prices in the United States, with the goal of estimating the degree to which each contributed to the doubling in turbine prices from 2002 through 2008, as well as the subsequent decline in prices through 2010 (our analysis does not extend into 2011 because several of these drivers are best gauged on a full-year basis due to seasonality issues). The first four of these drivers can be considered, at least to some degree, endogenous influences – i.e., those that are largely within the control of the wind industry – and include changes in:

1) *Labor costs,* which have historically risen during times of tight turbine supply;
2) *Warranty provisions,* which reflect technology performance and reliability, and are most often capitalized in turbine prices;

3) *Turbine manufacturer profitability*, which can impact turbine prices independently of costs; and
4) *Turbine design*, which for the purpose of this analysis is principally manifested through increased turbine size.

The other three drivers analyzed in this study can be considered exogenous influences, in that they can impact wind turbine costs but fall mostly outside of the direct control of the wind industry. These exogenous drivers include changes in:

1) *Raw materials prices,* which affect the cost of inputs to the manufacturing process;
2) *Energy prices,* which impact the cost of manufacturing and transporting turbines; and
3) *Foreign exchange rates,* which can impact the dollar amount paid for turbines and components imported into the United States.

The individual impacts of each of these seven drivers of wind turbine prices are summarized in Table ES-1. To focus attention on the overall trends, the table presents cumulative impacts over the two periods of major turbine price movements in the past decade – i.e., the doubling in turbine prices from 2002-2008, and the subsequent decline in prices through 2010.

Figure ES-2, meanwhile, shows yearly impacts, but of all seven drivers combined rather than individually. The aggregate impact (blue line with circle markers) is plotted against the empirical turbine price trend line from Figure ES-1 (red line with star markers). In aggregate, these seven drivers explain nearly $600/kW of the ~$750/kW increase in average turbine prices observed from 2002-2008, and nearly $90/kW of the ~$195/kW decrease in 2009 and 2010 (Table ES-1). From 2003 through 2010, the bottom-up analysis of these seven drivers explains between 68% and 89% (depending on the year) of the cumulative empirical price movements (Figure ES-2).

Nonetheless, Figure ES-2 shows a growing gap between modeled and empirical turbine prices starting in 2005 and increasing through 2007, at which point the gap remains more or less constant until eventually narrowing in 2010. Some portion of this wedge could potentially be explained by:

- Likely increasing labor costs and profitability among component suppliers beginning around 2005; as noted earlier and discussed further in Section 3.3, we quantified these two drivers for turbine manufacturers, but not for their component suppliers;

Table ES-1. Cumulative Impact During Period of Turbine Price Increase (2002-08) and Decrease (2009-10) (2010 $/kW)

	2002-2008		2009-2010	
Endogenous Drivers	*+376*		*-37*	
Labor Costs	+91		+12	
Warranty Provisions	+42		-20	
Profit Margins	+59		-78	
Turbine Scaling	+184		+50	
Exogenous Drivers	*+219*		*-53*	
Materials Prices	+71		-31	
Steel		*+65*		*-29*
Iron		*+7*		*-2*
Copper		*+9*		*+1*
Aluminum		*+2*		*-1*
Fiberglass		*-11*		*0*
Energy Prices	+12		-7	
Diesel		*+10*		*-4*
Coal		*0*		*0*
Natural Gas		*+2*		*-3*
Currency Movements	+136		-15	
Total Impact	*+595*		*-89*	

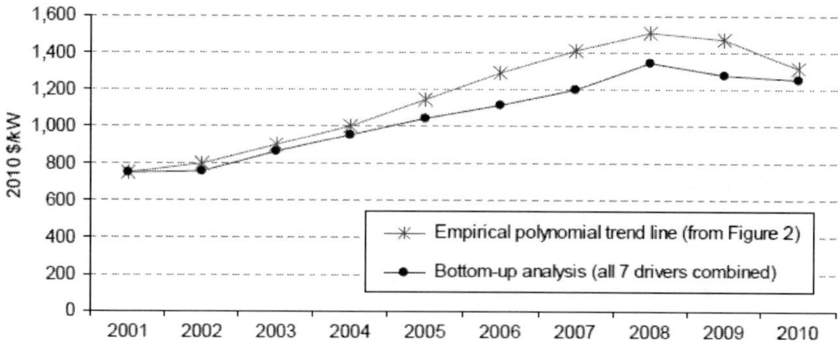

Figure ES-2. Yearly Impact of All Seven Drivers Combined vs. Empirical Price Curve.

- Higher exchange rate pass-through than the 50% assumed here (see Appendix C for more discussion of exchange rate pass-through);
- Turbine design and engineering improvements beyond the scaling effects analyzed in Section 3.4; and
- Methodological issues, such as the necessary reliance on certain data from Vestas when the market was supplied by multiple manufacturers (and especially GE Wind) over this period.
- Finally, it is clear from this analysis that there is no single, dominant factor that drove turbine prices higher from 2002-08, or that has yielded lower prices since that time. Turbine up-scaling is, by a significant margin, the largest single driver, although the estimated cost impact associated with up-scaling can be seen as a reasonable expense given the performance improvements garnered by larger turbines. Currency movements are also found to have played a sizable – though somewhat uncertain – role, as have changes in labor costs and material prices. Changes in manufacturer profit margins, warranty provisions, and energy prices are found to have played a less-significant, but non-negligible role.

1. INTRODUCTION

A considerable literature has developed using learning curve theory to explore how increases in cumulative wind power capacity (and other factors) have historically driven down wind energy costs (for a brief survey of the peer-reviewed literature, see Wiser et al. 2011a). The principal parameter calculated by these studies is the learning rate: for every doubling in cumulative production or installation, the learning rate specifies the associated percentage reduction in costs. Learning rates based on historical data are then often used to forecast future cost developments. As an example, Wiser and Bolinger (2011) calculate a learning rate of 14.4% for the installed cost of wind power projects in the United States during the period between 1982 and 2004, meaning that for each doubling in cumulative installed wind capacity worldwide over this period, installed wind project costs in the U.S. fell by 14.4% on average.

These historical cost reductions, in concert with governmental policies and other drivers, helped to fuel rapid growth in the industry, both domestically and abroad, starting around the turn of the century (Figure 1). In fact, although wind power technology has been commercially available for decades, more than 90% of all wind power capacity both in the US and worldwide has been installed in just the last 10 years. Over this period, global installed wind power capacity morethan-doubled in the four years from 2002 through 2005, and then again in the three years from 2006 through 2008; it is currently on track to double yet again by late 2011.

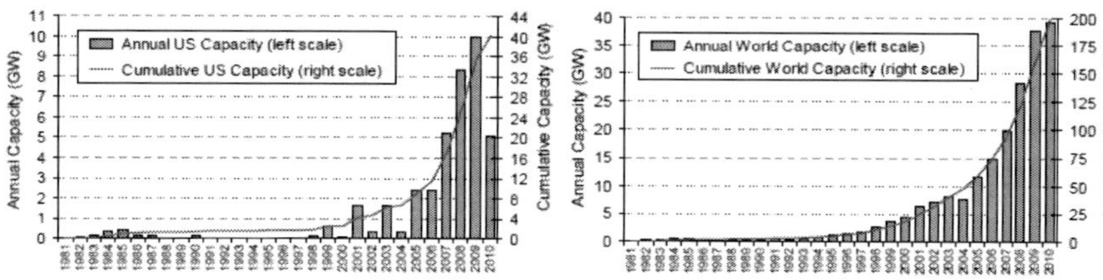

Figure 1. US and Global Installed Wind Power Capacity: 1981-2010.

Consistent with standard learning curve theory, the most-recent doubling expected by late 2011 has, in fact, been accompanied by significant cost reductions: as demonstrated later in Section 2, wind turbine prices in the U.S. have fallen somewhere on the order of 20%-33% on average since 2008. By some accounts, these turbine price declines, in combination with improvements in turbine design and performance, will result in a lower cost of wind electricity among projects currently being built than has ever before been possible (Wiser et al. 2011b).

It is important to recognize, however, that the substantial turbine price declines since 2008 started from elevated levels that, themselves, were not consistent with a simple understanding of standard learning curve theory. Rather than the nearly 30% decline in wind project costs that learning curve theory would have expected from 2002 through 2008 as a result of the two doublings in global installed capacity over this period, reported wind project costs in the U.S. actually increased by more than 50% percent over this period, due primarily to a doubling in wind turbine prices. This doubling in wind turbine prices through 2008, and the subsequent decline since then, mark a substantial divergence from the simple application of learning curves to cumulative wind power installations.

This divergence has important implications for the wind industry, policymakers, research and development (R&D) program managers, and energy analysts. With the wind industry only recently becoming a serious contributor to the power sector in the U.S. and globally, it must take care that unexpected cost inflation does not price wind out of the market, leading to demand destruction. Policymakers who count on wind to provide a growing share of the world's electricity needs – and who enact policies aimed to achieve that goal – want reassurances that wind can meet this challenge in a cost-effective manner (and that wind will eventually be able to wean itself off of direct public policy support altogether). R&D managers need to understand past cost trends in order to target future research most effectively. Finally, energy analysts who have heretofore placed some faith in the simple application of learning curves to project future technology costs must potentially reevaluate their beliefs and develop a more nuanced understanding of the drivers of wind (and other forms of) power costs.

Common to all four sets of stakeholders is a growing need to understand what specific factors – if not learning effects – have been driving recent wind power cost trends, and in particular the doubling in wind turbine prices from 2002 through 2008. This article seeks to contribute to such an understanding, with a specific focus on the cost of wind turbines deployed onshore in the United States. In doing so, it builds on the work of other studies that have begun to develop a deeper understanding of historical renewable energy cost drivers beyond simple, traditional concepts of learning (see, e.g., Nemet 2006; Papineau 2006, Ferioli et al. 2009, Yu et al. 2011), as well as those that have examined in some detail other causal influences to wind power costs, both on- and offshore (e.g., Bolinger and Wiser 2009, Milborrow 2008, Berry 2009, Blanco 2009, Greenacre et al. 2010, Carbon Trust 2008, Willow & Valpy 2011, BWEA & Garrad Hassan 2009, Ernst & Young 2009, Dinica 2011).[1]

To set the stage, Section 2 documents the increase in onshore wind turbine prices from 2002 through 2008 and the subsequent decline through 2010 using empirical data from the United States, as well as data provided by Vestas – the second-largest wind turbine supplier in the US market over this period. Section 3 examines seven different drivers that have been implicated to varying degrees in the run-up in wind turbine prices through 2008. The first four are, to a degree, endogenous influences that should be influenced by technology learning, and include labor costs, warranty provisions, turbine manufacturer profitability, and turbine design (namely, an increase in turbine nameplate capacity, hub height, and rotor diameter – all of which enable greater energy capture). The other three drivers are, to a greater extent, exogenous influences that fall outside of the scope of traditional learning curve theory, and include changes in raw materials prices, energy prices, and foreign exchange rates. Based on the analysis in Section 3, Section 4 presents the approximate degree to which each of these seven drivers, both individually and in aggregate, is found to have contributed to the overall movement in wind turbine prices over this period. Section 5 concludes by drawing insights from the analysis, and using them to look ahead to 2011 and beyond.

Before proceeding, we emphasize that this article focuses solely on wind turbine prices, rather than on the total installed cost of wind projects (which also includes balance of plant costs) or on the levelized cost of wind generation (which is further affected by financing terms, operating and maintenance expenses, and the amount of electricity generated). Wind turbine costs generally account for roughly 60%-70% of total installed project costs, and a slightly lower percentage of the levelized cost of wind generation (due to the latter also reflecting O&M and financing costs). Though it is ultimately the levelized cost of generation that is the most important of these three cost metrics, understanding trends in wind turbine pricing is a critical element to understanding trends in the levelized cost of generation.

2. Wind Turbine Price Trends in the United States

Berkeley Lab has gathered price data on 81 U.S. wind turbine transactions totaling 23,850 MW announced from 1997 through early 2011. Because of limitations in the data sources – most of which are press releases and news reports – the precise content of many of the individual turbine transactions is not known, though most transactions likely include only the turbines and towers delivered to the project site, as well as limited warranty and service agreements. Balance of plant ("BOP") construction, including foundation construction and turbine erection, is most often handled separately by engineering, procurement, and construction ("EPC") contractors (Bloomberg NEF 2011a, Fowler 2008).[2] Nevertheless, because of this uncertainty and the fact that our data sources are diverse, emphasis should be placed on overall trends in the data, rather than on individual data points.

Figure 2 depicts these reported wind turbine transaction prices (along with the associated trend line), broken out by the size of the transaction (in MW). Because visibility surrounding wind turbine transactions has declined in recent years,[3] Figure 2 also presents a range of reported pricing for transactions signed in 2010 and so far in 2011, sourced from Bloomberg NEF (2011b) as well as wind industry contacts. Finally, serving only as a quality check on our post-2004 transaction sample, Figure 2 includes average turbine prices reported by Vestas for the years 2005 through 2010 (Vestas 2011b, 2011c, 2011d) and converted into 2010 U.S. dollars.[4]

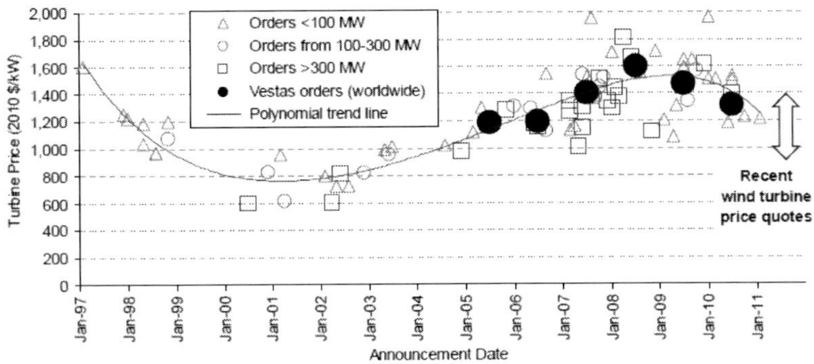

Source: Berkeley Lab, Vestas (2011b, 2011c, 2011d), Bloomberg NEF (2011b).

Figure 2. Wind Turbine Prices in the United States.

After hitting a low of roughly $750/kW from 2000 to 2002, average wind turbine prices *doubled* through 2008, rising to an average of roughly $1,500/kW. Wind turbine prices have since declined substantially, with price quotes for transactions executed in 2010 and to date in 2011 ranging from $900-$1,400/kW. These figures suggest price declines of as much as 33% or more since late 2008, with an average decline closer to perhaps 20% for orders announced in 2010 (as opposed to in 2011). As of July 2011, BNEF (2011c) was reporting U.S. transactions averaging $1,100/kW, though actual prices are highly dependent on the selected turbine design, with technology designed for lower wind speed sites (i.e., deploying higher hub heights and larger rotor diameters) coming in at higher pricing points.

Though other generation technologies experienced similar cost trends over this period,[5] these large swings in wind turbine prices merit an explanation. Moreover, the fact that the cumulative global installed capacity of wind power doubled *twice* during the period of rising prices suggests that we must look somewhere other than traditional learning curve theory for this explanation.

3. Wind Turbine Price Drivers

Taking a bottom-up approach, this section examines seven potential drivers of wind turbine prices in the United States, with the goal of estimating the degree to which each contributed to the increase in turbine prices from 2002 through 2008, as well as the subsequent decline in prices through 2010; the continued turbine price reductions experienced in the first half of 2011 are addressed to only a limited extent, in Section 5.[6] The first four of these drivers can be considered, to at least some degree, endogenous influences that should benefit from learning: these include labor costs (which are impacted by labor rates and quantities, the latter of which is endogenous), warranty provisions (which reflect technology performance and reliability), turbine manufacturer profitability (which tends to decline as industries grow and mature), and turbine design (for the purpose of this analysis, principally manifested through increased turbine size). The other three drivers analyzed in this study can be considered largely exogenous in that they can impact wind turbine costs but are not readily affected by learning in the wind industry, and include changes in the commodity price of raw materials and energy inputs to manufacturing processes, as well as movements in foreign exchange rates. Though the distinction between endogenous and exogenous influences is by no means absolute, and a number of the drivers analyzed here have elements of both,[7] it is nevertheless helpful in at least loosely understanding the relative importance of different categories of influences. It should also be noted that the seven specific drivers analyzed here do not fully explain the wind turbine price trends that have been empirically observed; other possible drivers are identified in the pages that follow, though available data do not allow a complete analysis of those additional drivers.

The next three sections (3.1-3.3) – on labor costs, warranty provisions, and turbine manufacturer profitability – draw heavily from the financial reports of one of the largest wind turbine manufacturers in the world – Vestas. Though it is not the intent of this report to focus exclusively on Vestas turbines or reach conclusions that are applicable only to Vestas, relying on Vestas' financial data nevertheless is both necessary and logical for reasons explained in the text box below.

Reliance on Vestas' Financial Reporting

During the period of interest to this study – 2002 through 2010 – GE (45%), Vestas (15%), Siemens (12%), and Mitsubishi (8%) together accounted for nearly 80% of all newly installed wind power capacity in the US. All four of these companies are publicly traded and are therefore required to disclose financial data, but only one of them – Vestas – is a "pure-play" wind turbine manufacturer.

The other three are heavy industrial conglomerates and do not disaggregate financial reporting for their wind turbine divisions in sufficient detail to be of use to this study. Gamesa (7%) and Suzlon (6%) together captured another 13% of the market over this period and are more wind-focused than the three large conglomerates, but nevertheless still have some financial reporting complexities related to subsidiaries (e.g., Gamesa's aeronautics and solar divisions in earlier years; Suzlon's investments in another turbine manufacturer – REpower – as well as Hansen transmission), and otherwise do not, in some cases, report relevant data back to 2001. In other words, among the six turbine manufactures that together accounted for 92% of all capacity installed in the U.S. from 2002-2010, only Vestas files financial reports that contain directly applicable and complete (notwithstanding recent accounting changes at Vestas, discussed in the text box on page 10) historical data on several drivers of interest to this study: labor costs, warranty provisions, and profitability.

Although we rely principally on Vestas for financial data by necessity, there are nevertheless several reasons to think that trends revealed through Vestas' filings can serve as a suitable proxy for the entire US market over this period – at least to the degree of precision required for this high-level analysis.

First, Vestas ranked second with 15% market share over this period (though admittedly far behind GE at 45%), which makes it a more representative choice than, for example, Gamesa or Suzlon; Vestas was also the largest manufacturer globally over this entire period, ensuring a degree of temporal stability in the resulting financial reports. Second, though turbine design variations existed among manufacturers over this period, the most widely deployed turbines were relatively homogenous, at least among the top U.S. turbine suppliers (e.g., Siemens is only now beginning to deploy direct-drive technology), which suggests that variations in cost structures as a result of fundamental design differences should be relatively modest. Third, because we are principally interested in *changes* in costs over time, rather than the absolute value of those costs, any differences in the underlying cost structure among manufacturers is less important than it might otherwise be. Fourth, the U.S. wind turbine market can be presumed to be competitive – e.g., witness how closely Vestas' reported average pricing matches the empirical price data presented earlier in Figure 2 – and competitive pressures should narrow the degree of variation in financial parameters impacting turbine suppliers over this period.

Although these four factors provide some comfort that Vestas can, indeed, serve as a suitable proxy for the entire U.S. market over the period of interest, in the limited cases where it is possible, we also present non-Vestas data to provide support for the trends revealed in Vestas' financial reports.

Finally, though not directly linked, the use of Vestas financial data also provides some continuity with the analyses of raw materials and energy costs presented later, which use data on Vestas turbines gleaned from several recent life-cycle analyses.

3.1. Labor Costs

Learning effects should theoretically drive down labor costs. As more and more technology is manufactured and deployed, manufacturers should become more efficient at utilizing – or even in some cases eliminating (e.g., through automation) – labor. When the pace of deployment is so rapid that it strains the supply of available labor, however, short-term cost increases are possible. Such was the case with the dramatic growth in installed wind power capacity over the past decade. As wind turbine and component manufacturers scrambled to address the scarcity of turbine supply, labor costs rose, leading to upward pressure on wind turbine prices.

Figure 3 shows Vestas' staff costs back to 2001, along with the cumulative change since then, expressed in 2010 \$/kW (delivered).[8] As shown, Vestas' reported labor costs increased by \$142/kW through 2009, with a subsequent \$39/kW decline in 2010. Though Vestas is understood to have historically had higher staff costs than some other turbine manufacturers (Efiong and Crispin 2007), the temporal changes in staff costs reported by Vestas are consistent with some other manufacturers. For example, though not shown in the figure, Suzlon's annual financial reports show a similar increase in labor costs since 2003 (available data from Suzlon do not go back any further): a \$115/kW increase through 2009, with a \$15/kW decline in 2010.

Not included here is an analysis of labor costs faced by component suppliers, for reasons discussed later in the turbine manufacturer profitability section (Section 3.3). However, if trends among component suppliers mirror those of the major turbine suppliers for which data could be obtained (Vestas and Suzlon), then the impact of labor costs on the overall turbine pricing trends presented in Section 2 would be even greater than estimated above.

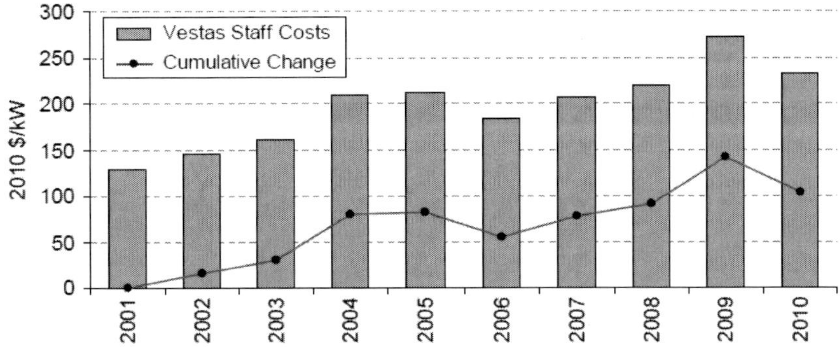

Source: Vestas annual reports for 2001-2010.

Figure 3. Vestas Staff Costs.

3.2. Warranty Provisions

In most cases, turbine prices include the expected cost (properly discounted) of standard warranty terms. In some cases, the expected cost of any extended warranties or service contracts entered into at the time of turbine sale may also be capitalized into the turbine price. In either instance, the amount set aside by turbine manufacturers to pay for future warranty obligations will depend in large part on how reliable a turbine is perceived to be, which in turn will reflect current and past operating experience with similar machines. Increases or decreases in actual warranty claims will eventually influence the size of the warranty provisions set aside for future claims, which in turn will impact wind turbine prices.

Figure 4 shows a synthetic time series of Vestas' warranty provisions back to 2001, along with the cumulative change in that series since then (the text box on page 10 explains why and how the synthetic time series was constructed).

The conversion to 2010 $/kW (delivered) was accomplished exactly as described earlier in footnote 8 for labor costs – i.e., by design, the impact of exchange rate movements over time are not captured.

The steep $78/kW increase in warranty provisions through 2005 reflects concerns throughout the industry in the early-to-mid-2000s about turbine reliability and warranty claims for both on- and offshore projects. For Vestas, the most visible manifestation of this concern (though involving an offshore project) was the well-publicized replacement of all 80 transformers and generators at the 160 MW Horns Reef wind project within just two years of its 2002 commissioning.

In late 2005, Vestas implemented the "Will to Win" campaign, which focused on returning the company to healthy profitability through a dual strategy of cost reduction *and* price increases (Vestas 2005).

One of the first acts of this campaign was to reduce the standard turbine warranty period to just two years (from as long as five years previously), which is reflected in the $48/kW drop in provisions in 2006. Since 2006, provisions have oscillated somewhat, but in 2010 took another concerted movement downward, as Vestas continued to progress on quality and reliability improvements through its "Six Sigma" program.[9] From 2001 to 2008, Vestas' warranty provisions increased by $42/kW; since 2008 they've dropped by $20/kW.

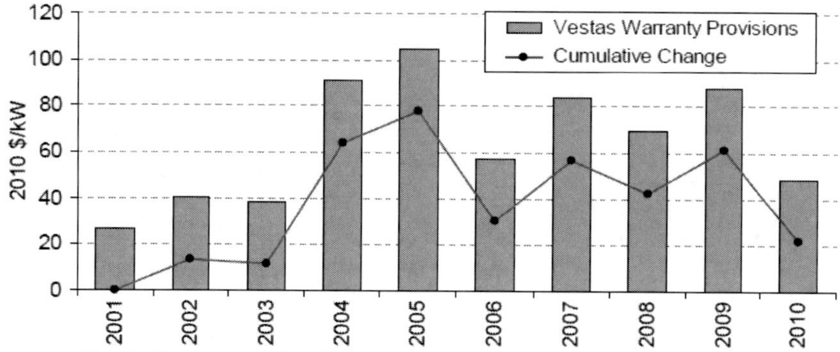

Source: Synthetic time series based on Vestas annual reports for 2001-2010 (methodology described in text box on page 10).

Figure 4. Vestas Warranty Provisions.

Though not shown in Figure 4, Gamesa's annual financial reports show a similar increase in warranty provisions (though for the entire company – not just Gamesa's wind division) since 2001: a $32/kW increase through 2008, followed by a $9/kW decline since then.

In addition, the unexpected increase in warranty expenditures in the mid-2000s, and the resultant increase in warranty provisions (at least through 2007), is further confirmed industry-wide by Efiong and Crispin (2007), who report warranty expenditures increasing from 3.5% of sales in 2004 to 4.75% in 2007.

Accounting Changes at Vestas

With its 2010 annual report, Vestas implemented a significant accounting change that impacts how revenues (which are closely linked to earnings and profitability, analyzed in Section 3.3) and warranty provisions (analyzed in Section 3.2) are accounted for. A review of these changes is warranted.

Revenues: Vestas previously accounted for revenue from its "supply and installation" contracts (which comprised 40% of new global orders in 2010) on a "percentage of completion" basis, which meant that significant revenue was realized at the time a turbine order shipped (i.e., a significant portion of Vestas' obligation under such contracts was fulfilled at the time of shipment).

Under the new accounting procedures, revenue from such contracts is realized only upon delivery of the project to the customer (i.e., upon the transfer of risk and benefits to the customer). This change has the impact of deferring revenue until a later date. Revenue recognition from "supply-only" and "EPC/Turnkey" contracts has not changed: the former has always occurred at the time of delivery, while the latter will continue to occur on a "percentage of completion" basis.

Warranty Provisions: In addition to offering standard warranties for its turbines, Vestas has also in recent years marketed long-term service contracts that cover routine maintenance and expected repairs over terms as long as 10 years. Although these service contracts include warranty-like obligations, Vestas previously accounted for such obligations in real time – i.e., in the year in which they actually became expenses. With the accounting change, Vestas now accounts for expected future repairs under service contracts by setting aside warranty provisions up front, in the year in which the service contract is signed. For example, if Vestas expects to replace a gearbox in year 7 of a 10-year service contract, previously the cost of that gearbox replacement would have been realized in year 7 (i.e., when it actually occurred), whereas now Vestas sets aside sufficient provisions in year 1 to pay for the expected eventual replacement. This change also has the effect of boosting future profitability at the expense of current profitability.

With these accounting changes, Vestas re-stated its income statement going back only five years (to 2006). Our analysis of warranty provisions and profit margins (labor costs were not affected by these accounting changes), however, extends further back in time – i.e., ideally back to 2001.

To solve this mis-match, an indexing approach was used to create a synthetic time series that reflects the accounting change from 2006-2010 yet also provides the history we need over the entire period. The index starts with the old (pre-restatement) numbers from 2001 through 2006; in 2007 and all later years, that year's annual change in the new (restated) numbers is added to the previous year's index value to continue the time series. The result is a "synthetic" time series that, necessarily, will not exactly match Vestas' most-recent annual report, but that nonetheless effectively captures temporal trends in both warranty provisions and manufacturer profitability.

3.3. Turbine Manufacturer Profitability

Wind turbine manufacturers must be profitable to remain in business. In the face of rising costs, profitability can only be maintained by raising turbine prices commensurately – these cost-related price increases are quantified in other sections of this report that focus on labor, warranty, raw materials, and energy costs (and, more subtly, foreign exchange rate movements). In contrast, this section focuses on changes in profitability itself (i.e., independent of costs). For example, the "seller's market" for wind turbines that developed starting in 2005 provided an opportunity for wind turbine manufacturers to increase their profitability by raising prices above and beyond the amount required to cover rising costs. Conversely, the dwindling number of and intense competition for new orders since the global financial crisis of 2008/2009 has led to price reductions, partially achieved through lower profit margins. This section estimates changes in wind turbine prices attributable to changes in turbine manufacturer profitability.

Figure 5 shows a synthetic time series of Vestas' operating profit (i.e., earnings before interest and taxes, or "EBIT") back to 2001, along with the cumulative change in that series since then (the text box on page 10 explains why and how the synthetic time series was constructed). The conversion to 2010 $/kW (delivered) was accomplished as described earlier in footnote 8 for labor costs – i.e., by design, exchange rate movements over time are not captured, since they are analyzed separately in Section 3.7.

After its per-kW profit margin steadily declined from 2001 through 2005, Vestas' "Will to Win" campaign, launched in late 2005, focused on returning the company to healthy profitability through a strategy of cost reduction *and* price increases (Vestas 2005). Perhaps as a result, Vestas' profits rose sharply in 2006, and then again in 2008, before falling in the wake of the global financial crisis and resulting softness in turbine sales. In aggregate since 2001, Vestas' operating profit rose \$59/kW through 2008, and then fell by \$78/kW through 2010. The overall average increase in turbine manufacturer profitability from the widely-understood-to-be depressed levels of the early 2000s (Milborrow 2008) until 2008 is confirmed industry-wide by Efiong and Crispin (2007), who show EBIT margins of -8% in 2004 increasing to 2% in 2005 and then to 6-7% in 2006-2007; Bloomberg NEF (2009), meanwhile, reports lower profit margins after the financial crisis at the end of 2008.

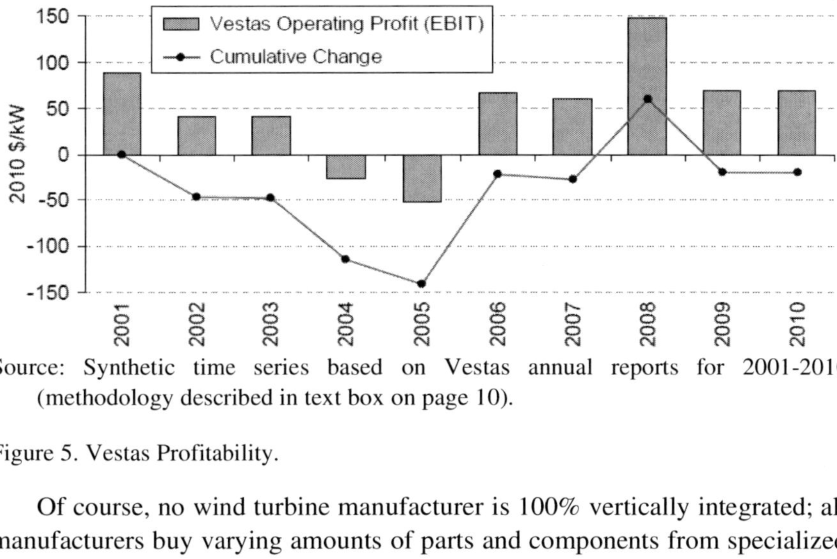

Source: Synthetic time series based on Vestas annual reports for 2001-2010 (methodology described in text box on page 10).

Figure 5. Vestas Profitability.

Of course, no wind turbine manufacturer is 100% vertically integrated; all manufacturers buy varying amounts of parts and components from specialized suppliers. As such, it is not just the profitability of Vestas (or any other wind turbine manufacturer) that influences wind turbine prices; the profitability (and labor costs, discussed earlier) of turbine component suppliers will also have an impact. Quantifying component supplier profitability (and labor costs) is challenging, however, because many component suppliers are not publicly traded (which limits the type of financial information they are required to disclose), and those that are often manufacture many other goods besides wind turbine parts (making it difficult to isolate wind-related impacts). Even if a representative data sample were available, it is not clear how one would go about quantifying impacts on a \$/kW basis.[10] For these reasons, our analysis does not include the impact of component supplier profitability (and labor costs), while acknowledging that such impacts do exist and have likely exacerbated the trends presented above for Vestas. Bloomberg NEF (2009), for example, reports that in the wake of the financial crisis, that both turbine and component manufacturers have experienced lower EBIT margins.

3.4. Increasing Turbine Size and Energy Capture

The average nameplate capacity of wind turbines installed in the U.S. doubled over the period of interest, from just under 0.9 MW in 2001/2002 to nearly 1.8 MW in 2010 (Figure 6). Along with this doubling in capacity, the average turbine hub height and rotor diameter also increased: hub height by one-third (from just under 60 meters to 80 meters) and rotor diameter by nearly two-thirds (from just over 50 meters to nearly 85 meters). Because mass scales more rapidly than height or length – e.g., taller towers are not only taller, but also need to be wider and thicker (and therefore heavier) to support the extra height – the rapid growth in turbine size has also impacted wind turbine prices on a $/kW basis. The fact that the capital cost of turbines can increase with size is widely understood (e.g., Dinica 2011, EWEA 2009), but the advantages in terms of lower balance of plant costs and higher levels of energy production typically outweigh those turbine price increases.

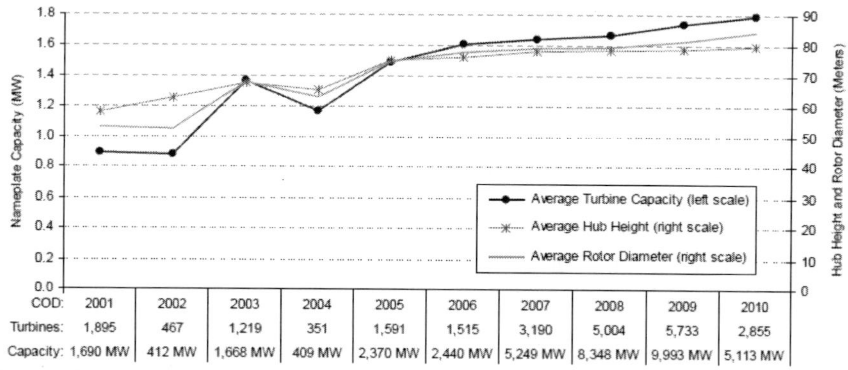

Source: Berkeley Lab Database.

Figure 6. Average Turbine Capacity, Hub Height, and Rotor Diameter.

To analyze the impact of turbine up-scaling, we use the wind turbine cost and scaling model developed at the National Renewable Energy Laboratory ("NREL") and described in Fingersh et al. (2006). The model consists of a series of cost functions developed in the early-to-mid-2000s and based on a

1.5 MW turbine platform – i.e., more or less right in the middle of both our time period of interest *and* turbine size range.

The functions calculate how both mass and costs might change in response to changes in certain turbine design parameters such as nameplate capacity, hub height, and rotor diameter. The model also estimates how turbine scaling affects the levelized cost of electricity ("LCOE") generated by the turbine, based on general wind shear assumptions, power curve data, and other parameters.

Figures 7 and 8 provide the results of simply inputting each combination of average nameplate capacity, hub height, and rotor diameter from Figure 6 into the NREL scaling model.[11]

As turbine capacity and size scales up, so does mass, which in turn increases the estimated cost of the manufacture and transport of the turbine – by roughly $230/kW since 2001 (Figure 7).[12]

As noted earlier, however, this capital cost increase is not without benefit – Figure 8 shows that larger and taller turbines deliver higher capacity factors and a lower levelized cost of electricity.[13]

In other words, unlike most of the other factors that have driven turbine prices higher over the period of interest, turbine up-scaling is the result of a conscious decision to design larger and more powerful turbines that can capture more of the wind's energy and convert it to electricity at lower costs. In this case, the benefit (lower LCOE) outweighs the incremental cost (higher $/kW turbine price) of scaling.

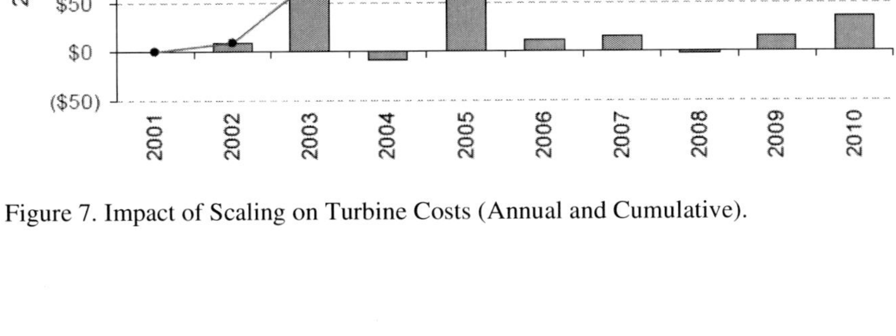

Figure 7. Impact of Scaling on Turbine Costs (Annual and Cumulative).

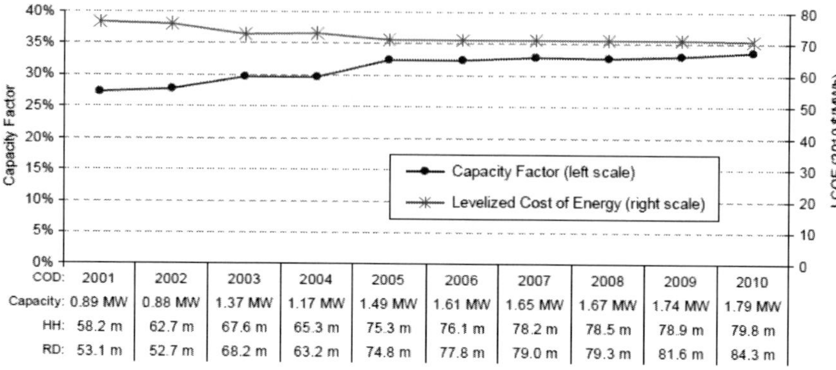

Figure 8. Results from NREL Turbine Scaling Model: Capacity Factor and LCOE.

This analysis suggests that turbine scaling has increased turbine pricing over the 2002 through 2010 period, but with the benefit of increased energy capture and lower LCOE. Two caveats, however, suggest that turbine design parameters are not fully captured in the analysis presented here. First, the NREL cost model relies on standard relationships between component size, outside the bounds of these standard relationships, actual scaling-related cost influences may have differed from what is presented above. Second, the analysis above was focused exclusively on turbine scaling, but turbines today also have features and capabilities that exceed those that were available in the early 2000s (e.g., sophisticated control systems, grid-friendly features, etc.). These additional design and engineering improvements – many of which were likely implemented to boost reliability in the face of the rising warranty claims/provisions shown earlier in Section 3.2 – impose an additional cost (and benefit) that is not captured by our focus on scaling alone.

3.5. Raw Materials Prices

In addition to the four somewhat-endogenous influences on wind turbine prices explored in the previous sections, there are also three exogenous influences that, although they can be managed to an extent, are nevertheless largely out of the direct control of the wind power industry. These include changes in raw materials prices, energy prices, and foreign exchange rates – each of which will be analyzed in turn in the next three sections (3.5-3.7).

Wind turbines are material-intensive. Each individual tower, perhaps extending eighty meters or higher, may require several hundred metric tonnes of steel. Significant amounts of fiberglass and resin are required for the blades and nacelle. Iron, copper, and aluminum are also commonly used for various elements of the drive train and generator. Sustained movements in the price of any of these raw materials could explain part of the increase and subsequent decrease in turbine prices over the past decade, just as similar trends impact the cost of other forms of power generation equipment (e.g., Chupka and Basheda 2007). Using somewhat different approaches and data sources, previous efforts to estimate these impacts on wind energy include Greenacre et al. (2010) and Milborrow (2008).

Table 1 provides a breakdown, by mass and percentage of total mass, of the five major materials used to manufacture four different wind turbine models – three from Vestas and one from Gamesa – which span a significant range of nameplate capacity (from 1.65 MW to 3.0 MW), rotor diameter (from 80 to 112 meters), and tower height (from 67 to 84 meters). The data are compiled from life-cycle analyses ("LCAs") of wind power projects using these four turbine types. Although these LCAs often provide more detail than reported here, Table 1 consolidates closely related materials (e.g., different types of steel) into single categories where logical, while ignoring other minor (at least in terms of mass) materials whose costs are difficult to track (e.g., electronics). In all cases, materials used in the foundations of these turbines (e.g., steel rebar and concrete) are excluded from Table 1, since the empirical turbine prices shown earlier in Figure 2 are generally assumed to *not* include foundation costs, which, again, most often fall under BOP or EPC contracts and are not covered in the present study.

The bottom of Table 1, which presents the mass of each of the five materials categories expressed as a percentage of total turbine mass, reveals two important findings that enable us to simplify the present analysis. First, the fact that these five materials account for more than 98% of the total mass of each of these four turbines suggests that focusing the analysis on just these five materials, rather than on the more-detailed breakdown of materials that can be found in some LCAs, is sufficient.[14] Second, the fact that the distribution of mass among these five materials is relatively consistent across the four different turbine models suggests that the analysis will yield similar results regardless of which turbine is analyzed. As a result, for reasons explained in the text box on page 17, the analysis of raw materials (and later energy) prices focuses exclusively on the Vestas V82 1.65 MW turbine (i.e., the first data column of Table 1).

Table 1. Condensed Bill of Materials for Four Different Turbines

Turbine make/model:	*Vestas V82*	*Gamesa G8X*	*Vestas V80*	*Vestas V112*
Nameplate capacity:	1.65 MW	2.0 MW	2.0 MW	3.0 MW
Tower height:	78 meters	67 meters	78 meters	84 meters
Rotor diameter:	82 meters	80 meters	80 meters	112 meters
Mass (kg per kW)				
Steel	96.3	82.3	104.7	81.7
Fiberglass/Resin/Plastic	18.2	11.1	12.3	16.3
Iron/Cast Iron	17.8	16.3	10.3	21.9
Copper	1.8	1.8	1.4	1.6
Aluminum	1.9	0.0	0.8	1.1
Total	*135.9*	*111.4*	*129.6*	*122.7*
% of Total Turbine Mass				
Steel	70%	74%	81%	66%
Fiberglass/Resin/Plastic	13%	10%	9%	13%
Iron/Cast Iron	13%	15%	8%	18%
Copper	1%	2%	1%	1%
Aluminum	1%	0%	1%	1%
Total	*98.3%*	*100.0%*	*99.9%*	*99.0%*

Source: Vestas 2006b, Martinez et al. 2009, Elsam 2004, D'Souza et al. 2011.

Figure 9 shows the cumulative percentage change in prices (in real 2010 dollar terms) since December 2001 for each of the five materials listed in Table 1 (details on how the individual time series price data for each material were sourced and constructed can be found in Appendix A). With the exception of fiberglass/resin, the prices for all other materials shown in Figure 9 have escalated at rates faster than the general rate of inflation over this period. Specifically, the four metals experienced massive price increases that uniformly peaked in mid-2008, before plunging sharply through the end of that year as a result of the global financial crisis. Notably, metals prices have since moved higher once again.

Assuming full pass-through of the commodity price changes shown in Figure 9 to the price of materials used to manufacture the Vestas V82 1.65 MW turbine,[15] the run-up in raw materials prices from 2002 through 2008 is estimated to have added roughly $71/kW to wind turbine prices (Figure 10). A significant portion of this increase ($47/kW) was erased in 2009 as metals prices plunged, but the recovery in prices in 2010 added back $16/kW. Among the five materials analyzed, the price of steel has had by far the largest impact on turbine prices (during both the period of increase and decrease), reflecting the relative importance of steel in terms of both mass and turbine price contribution (Table 5 in Section 4 shows results by individual material).

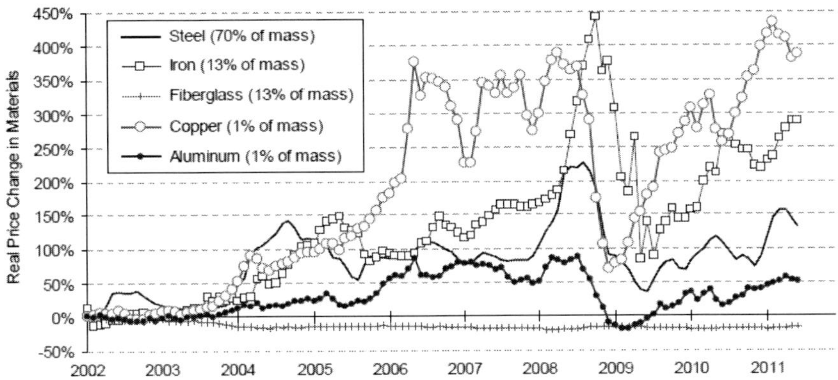

Figure 9. Cumulative Price Change in Raw Materials Since December 2001.

Choice of Vestas V82 1.65 MW LCA

As shown earlier in Figure 6, the average turbine size installed in the U.S. increased from roughly 0.9 MW in 2001 to 1.8 MW in 2010. This average turbine size range suggests that the Vestas V82, rated at 1.65 MW, is a more appropriate choice for analysis than any of the other three turbines featured in Table 1, each of which are larger 2 or 3 MW turbines. Because it falls within the average size range of turbines installed during this period, the 1.65 MW turbine minimizes distortions that could originate from analyzing the impact of commodity price changes independently from the impact of turbine scaling (when in reality, the two are somewhat interdependent – this issue is described more fully in footnote 11).

The choice of the Vestas 1.65 MW turbine is bolstered by further analysis of the data behind the average turbine sizes reported in Figure 6. These data reveal that Vestas installed 1,780 turbines rated at 1.65 MW in the U.S. between 1999 and 2010, while NEG Micon – which later merged with Vestas – installed another 60 such turbines in 2003 and 2004. In other words, not only does the 1.65 MW turbine more closely match the average turbine size over the period of interest, but Vestas has also installed a substantial number of 1.65 MW turbines in the U.S. (more than it has any other size turbine), distributed over the entire period of interest.

In contrast, through the end of 2010, Vestas had installed only twenty-two 2 MW turbine in the U.S. (all in 2009), and 323 turbines rated at 3 MW (the bulk of which were installed after 2007, and all of which are the V90 model, rather than the V112 shown in Table 1). Meanwhile, nearly 1,200 Gamesa 2 MW turbines have been installed in the U.S. through 2010, but only since 2005 (i.e., part way into the analysis period of 2002-2010).

Apart from size and distribution considerations, there is another reason to favor the Vestas 1.65 MW turbine for this analysis. The Vestas 1.65 MW LCA provides one of the most useful breakdowns of energy consumption during the manufacturing and transport phases. Most other LCAs aggregate energy consumption across different life-cycle phases (e.g., manufacture, transport, operation, disposal) in such a way that it is difficult or impossible to focus on just the manufacturing and transport phases – i.e., the two phases that impact turbine prices.

The cost of energy consumption is analyzed separately in Section 3.6. Based on these considerations, the analysis of raw materials (and later energy) prices focuses exclusively on the Vestas V82 1.65 MW turbine. Once again, though, the distributional uniformity across turbines in Table 1 indicates that results would not vary significantly if any of the other three turbines were instead chosen.

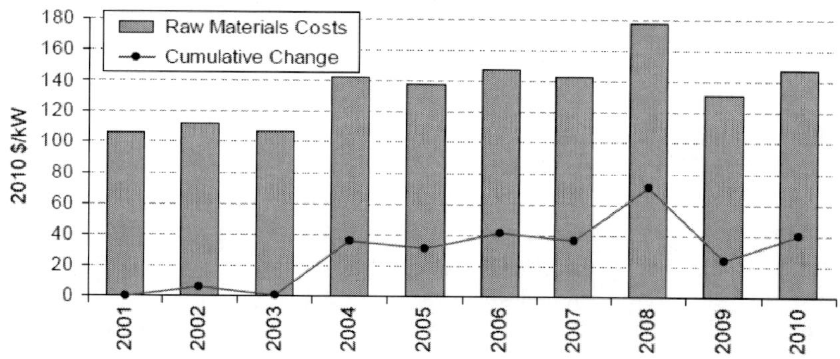

Figure 10. Raw Materials Costs.

3.6. Energy Prices

Life-cycle analyses of wind projects find that the amount of electricity generated by the project during its operating phase far outweighs the amount of energy consumed by all other life-cycle phases combined (for a literature survey, see Wiser et al. 2011). In most cases, the "energy payback period" is found to be less than a year. Nevertheless, it does take a significant amount of energy to manufacture a wind turbine and transport it – often over long distances – to a project site, and a sharp rise in the cost of energy used during these two phases could, therefore, have an impact on wind turbine prices. This section explores this possibility.

As it was in the previous section, Vestas' LCA of its V82 1.65 MW turbine (Vestas 2006b) is also the principal source used here for the amount of energy consumed during the manufacture and transport of a wind turbine. Considerable effort was required, however, to overcome two limitations in what is presented in the LCA.

First, assumptions were required to narrow the scope from the amount of energy consumed over all four life-cycle stages (manufacturing, transport, operation, and disposal) to just those two stages that impact turbine prices – i.e., manufacturing and transport.[16] Second, external sources of data were needed to help strip out the energy "embodied" in the materials themselves, as the cost of this embodied energy is already reflected in the price of the raw materials – changes in which were analyzed separately in the previous section (Section 3.5).

Table 7 of Vestas (2006b) shows the consumption of a variety of primary energy sources – including hard coal, lignite (soft coal), crude oil, natural gas, uranium, and a variety of renewable fuels (biomass, wind, hydro, solar, geothermal) – over the life of a V82 wind turbine, but does not break out energy consumption by life-cycle stage. Table 6 of Vestas (2006b) does provide such a breakout, but only for the fossil fuels (for reference, Tables 6 and 7 of Vestas (2006b) are reproduced here in Appendix B). If one assumes, however, that all of the nuclear and renewable fuels listed in Table 7 (of Vestas 2006b) are consumed during the manufacturing stage,[17] then one can use the numbers provided in that table for these resources, while using fossil fuel consumption from just the "Production" and "Transport" columns of Table 6 (of Vestas 2006b). Adding all of these together suggests that the amount of primary energy consumed during just the manufacturing and transport stages of the turbine's life cycle comes to 9.36 GJ/kW in total (Table 2).[18]

One more transformation is then required for the present analysis. Like most life-cycle analyses, the Vestas V82 LCA tracks the amount of *primary energy* consumed by the turbine over its lifetime.

Table 2. Primary Energy Consumed by Production and Transport of Vestas 1.65 MW Turbine

	GJ / kW	% Contribution
Crude oil*	2.635	28%
Hard coal*	3.808	41%
Lignite*	0.170	2%
Natural gas*	1.750	19%
Uranium	0.428	5%
Biomass	0.201	2%
Wind	0.008	0%
Hydro	0.348	4%
Solar	0.007	0%
Geothermal	0.002	0%
Total	9.357	100%

* Fossil fuels are a combination of "Production" and "Transport" columns from Table 6 of Vestas (2006b), converted to MJ/kWh using conversion factors in Table 7 of Vestas (2006b). All other fuels are pulled directly from Table 7 of Vestas (2006b).

Primary energy is the most basic and comprehensive expression of energy needs, and includes not only the energy consumed during the manufacturing process (for example), but also the energy embodied in the manufacturing inputs themselves (e.g., the energy required to convert iron ore into usable steel). Although this focus on primary energy is appropriate for calculating the turbine's lifetime energy payback (i.e., one of the goals of most LCAs), it is problematic for the present purposes, in that the materials prices analyzed in the previous section already include the cost of the embodied energy used to produce them. To avoid double-counting, embodied energy must therefore be removed from total primary energy consumption.

Table 1 of Schleisner (2000) provides an estimate of the amount of primary energy required to produce, transport, and manufacture a single kilogram of a variety of materials, including most of the materials used to build a wind turbine. Specifically, all materials listed in Table 1 of the previous section – with the exception of fiberglass/resin, for which "plastic (polyester and epoxy)" is substituted – are included in Schleisner (2000). Multiplying the mass of the five materials listed in Table 1 of the previous section – plus concrete19 – by the primary embodiedenergy content from Schleisner (2000) yields the values shown above in Table 3. In total, the materials used to produce a wind turbine are thereby estimated to contain roughly 6.372 GJ/kW of embodied energy, with steel and concrete together containing nearly 75% of the total.

Table 3. Primary Energy Embodied in Materials Used to Build a Wind Turbine

	Vestas 1.65 MW	Primary Energy Consumption	
	kg/kW	MJ/kg	GJ/kW
Steel*	112.7	25.65**	2.890
Concrete	487.9	3.68	1.795
Fiberglass/Resin	18.2	45.7***	0.831
Iron/Cast Iron	17.8	36.3	0.645
Copper	1.8	78.2	0.137
Aluminum	1.9	39.15**	0.074
Total	640.1		6.372

*Includes steel used in turbine foundation; hence, this mass does not match that shown earlier in Table 1, which excluded foundations.

**The primary energy content of steel and aluminum represent the average of the minimum and maximum values provided by Schleisner (2000).

*** Schleisner (2000) does not include fiberglass, so the energy content provided for "Plastic (polyester and epoxy)" is used instead.

Subtracting the amount of primary energy embodied in wind turbine materials (6.37 GJ/kW) from the total amount of primary energy consumed during the manufacturing and transport phases (9.36 GJ/kW) leaves just the amount of energy required to manufacture and transport the turbine itself: 2.99 GJ/kW. In other words, the energy consumption that is of relevance to this study is about one-third of the total primary energy required to manufacture and transport a wind turbine; the remainder is energy embodied in the materials themselves.[20]

Table 2 above lists the estimated percentage contribution of the various energy sources that, at least in the Vestas 1.65 MW LCA, supply the primary energy consumed by the manufacture, transport, and installation of a wind turbine. These proportions cannot be directly used in our analysis, however, because they include the energy embodied in materials. Assuming that a disproportionate amount of the embodied energy derived in Table 3 comes from coal (Schleisner 2000), we make a simplifying assumption that 60% of primary energy consumption (after subtracting out embodied energy) comes from the three fossil fuels (split evenly between them, and with oil taking the form of diesel fuel), with the remaining 40% coming from largely stable-priced energy sources (e.g., nuclear and renewables) – i.e., we assume that only 60% of energy consumption is subject to fuel price risk. Though approximate, these round proportions are broadly consistent with the Vestas 1.65 MW LCA (once coal-heavy embodied energy is stripped out), as well as the more-recent Vestas V112 LCA (D'Souza et al. 2011). Detailed "end-use" energy consumption data reported in the "Sustainability" section of Vestas' web site also broadly support this apportionment.

Figure 11 shows the cumulative percentage change in prices (in real 2010 dollar terms) since December 2001 for each of the three variable-priced energy sources (details on how the individual time series price data for each energy source were sourced and constructed can be found in Appendix A). All three fossil fuels have escalated at rates faster than the general rate of inflation over this period. Diesel fuel and natural gas prices experienced the largest increases through mid-2008, followed by a sharp reversal through early 2009 for diesel (but rising since then), and still continuing for natural gas (at least in the U.S., where increasing shale gas production has boosted supply). In comparison, the rise in coal prices has been slower and steadier.

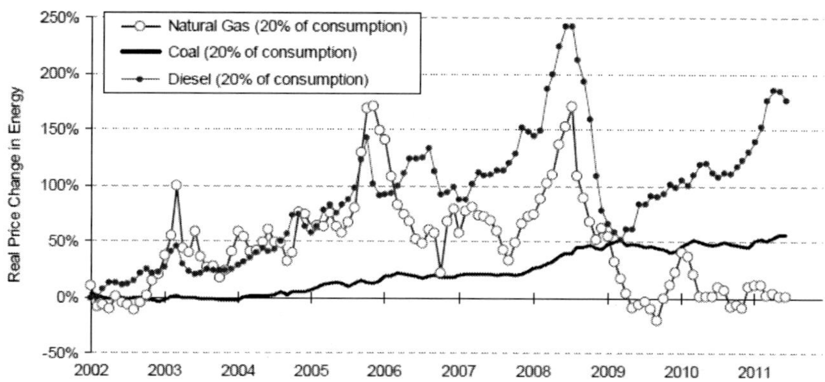

Figure 11. Cumulative Price Change in Energy Resources Since December 2001.

Despite a near-doubling in aggregate fossil energy prices from December 2001 through June 2008, the contribution of energy costs to the run-up in turbine prices over this same period appears to have been relatively modest, at roughly $12/kW (Figure 12). Moreover, with the sharp decline in energy prices since June 2008, more than half of this contribution – $7/kW – has been erased.

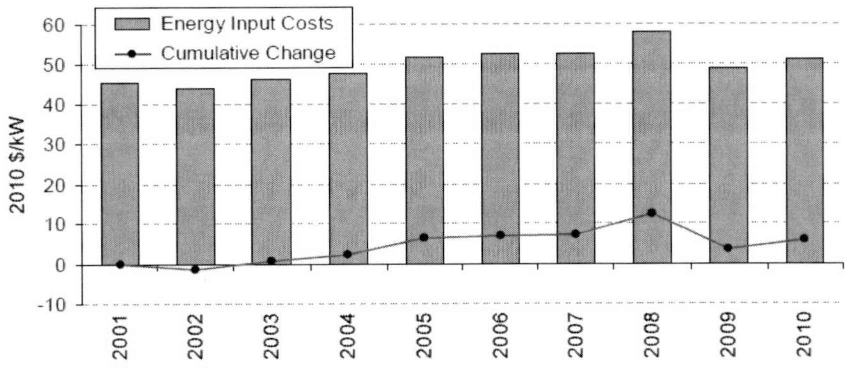

Figure 12. Energy Input Costs.

3.7. Foreign Exchange Rates

Historically, a substantial fraction of the wind turbines installed in the U.S. have been imported, either in part or in full, from other countries. As a result, the strength of the U.S. dollar ("USD") relative to the home currencies of turbine-exporting countries has impacted the USDdenominated price of wind turbines over time. All else equal, the USD-denominated price of wind turbines will increase as the USD loses value, and will fall as the USD strengthens. With the USD suffering from an extended period of weakness during the years of interest to this study, exchange rate impacts could explain a significant amount of the run-up in wind turbine prices in the U.S. over this period.

Although modern wind power technology can trace many of its roots back to turbines built and installed in California in the late 1970s and early 1980s, U.S. wind turbine manufacturing largely stagnated – along with the U.S. market for turbines – once California state and federal incentives were eliminated in the mid-1980s. It wasn't until the turn of the century that demand for wind power in the U.S. began to pick up again (Figure 1), and by that time, several European turbine manufacturers – who, after the collapse of the California market in the late-1980s, had been supplying growing markets in northern Europe throughout the 1990s – dominated the global wind turbine market, along with newly emerging competition from a number of Asian countries, most notably Japan and India. Although the U.S. never fully conceded wind turbine manufacturing to overseas companies – e.g., since buying Enron

Wind's assets in 2002,[21] U.S.- based GE Wind has become a leading supplier of wind turbines worldwide – it is nevertheless the case that the U.S. has been and remains one of the world's largest importers of wind turbines.

This picture, however, has begun to change in recent years. In addition to GE Wind, new U.S.- based wind turbine manufacturers – including Clipper Windpower – have more recently entered the market, primarily supplying U.S. projects. Foreign-based turbine manufacturers have also been investing in the U.S.: by the end of 2010, nine of the eleven wind turbine manufacturers with the largest share of the U.S. market in 2010 had one or more manufacturing facilities operating in the U.S. Finally, both domestic and foreign component manufacturers have also increasingly invested in U.S.-based manufacturing facilities (Wiser and Bolinger 2011).

As a result of this activity, the proportion of wind turbines and components that are imported into the United States has declined significantly in recent years. Based on trade data from the U.S. Department of Commerce, Wiser and Bolinger (2011) estimate that the fraction of installed wind turbine equipment costs imported into the U.S. fell from about 65% in 2005-2006 to 58% in 2007-2008, and then further to 40% in 2009-2010 (Figure 13). This most recent empirical estimate – which suggests roughly 60% local content in the US on average – is generally consistent with, or even conservative relative to, similar estimates released by individual wind turbine manufacturers: Gamesa claims to offer 60% local content in the US (Gamesa 2011), Acciona 77% (Waggoner 2011), and Vestas 80-90% (Vestas 2011a).

Even though it has subsided in recent years as the import fraction has declined and the degree of local content has increased, the risk of adverse exchange rate movements inflating the cost of wind turbines imported into the U.S. has been very real throughout the period of interest to this study (i.e., 2002 – 2010). Table 4 lists those countries that have accounted for the largest estimated share of wind turbine equipment imported into the U.S. over time,[22] and thereby make up the bulk of foreign currency exposure. Figure 14 shows that from the start of 2002 through the first half of 2008, the USD had lost at least 15% of its value relative to all but one of these currencies (the Mexican Peso). Of particular note, given the high concentration of wind turbine imports from Europe, USD weakness has been especially pronounced relative to European currencies: from the start of 2002 through mid-2008, the USD lost 40% of its value against the Euro (as well as the Danish Kroner, which has moved in lock-step with the Euro).

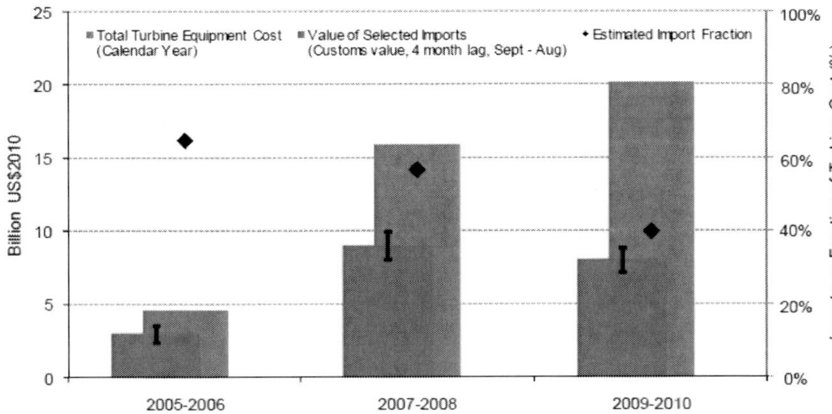

Source: Wiser and Bolinger (2011).

Figure 13. Estimated Wind Turbine Equipment Imports as a Fraction of Total Turbine Cost.

Table 4. Share of U.S. Wind Turbine Equipment Imports Over Time

	2002	2003	2004	2005	2006	2007	2008	2009	2010
Denmark	72.8%	54.9%	36.6%	62.1%	49.0%	37.1%	22.0%	28.1%	42.0%
Euro zone	0.0%	10.2%	0.7%	8.5%	18.6%	24.3%	30.5%	19.1%	8.7%
U.K.	0.0%	0.0%	2.9%	4.0%	2.8%	4.6%	4.1%	4.3%	0.1%
Japan	6.0%	21.6%	23.5%	10.8%	6.7%	10.5%	11.1%	20.3%	0.7%
India	0.9%	4.0%	0.0%	2.2%	15.2%	9.4%	5.9%	9.8%	16.3%
China	0.0%	0.1%	0.2%	0.3%	1.2%	5.0%	5.8%	5.9%	5.9%
Mexico	17.3%	3.6%	4.6%	2.0%	1.2%	1.4%	1.7%	1.5%	5.6%
South Korea	0.1%	4.6%	9.5%	1.7%	1.2%	2.6%	6.0%	5.3%	2.1%
Canada	1.8%	0.8%	21.7%	3.3%	0.8%	2.0%	4.3%	1.1%	8.4%
Other	*1.1%*	*0.2%*	*0.3%*	*5.1%*	*3.3%*	*3.1%*	*8.5%*	*4.6%*	*10.1%*
Total	100%	100%	100%	100%	100%	100%	100%	100%	100%

Source: LBNL analysis of U.S. Department of Commerce data, via US ITC DataWeb

All else equal, this prolonged period of USD weakness has likely inflated the USD-denominated cost of wind turbines imported into the U.S. The magnitude of the impact, however, is less certain. Assessing the impact of exchange rate movements on USD-denominated wind turbine prices over time is complicated by the fact that not only has the foreign content of turbines installed in the U.S. declined in recent years (Figure 13), but the mix of countries from which that declining foreign content originates has shifted over time as well (Table 4).

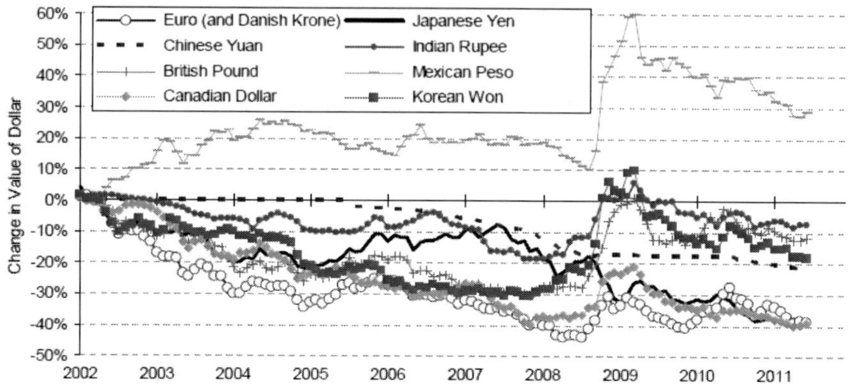

Figure 14. Cumulative Change in the Value of the Dollar Since December 2001.

To simultaneously account for these two moving pieces, we start with declining turbine equipment import fractions loosely based on Figure 13 above, but extrapolated back in time: 70% in 2002, 68.5% in 2003, 67% in 2004, 65% in 2005, and then dropping by five percentage points each year until reaching 40% in 2010. Next, we multiply these import fractions by the corresponding average turbine price for each year as reported earlier in Figure 2, to arrive at the total average dollar amount of the turbine price that could be subject to exchange rate risk in each year. Each year, we then apportion this "at risk" USD amount among the various foreign currencies shown in Table 4 for that year (grossing them up slightly so that they add to 100%), and then convert the resulting USD amounts into the corresponding foreign currency amounts using average annual exchange rates for that year. Next, we convert these foreign currency amounts back to USD using the average annual exchange rates for the *subsequent* year (i.e., we measure the impact of exchange rate movements over the interim one-year period), sum them, and then subtract the resulting USD sum from the starting USD amount. Finally, we multiply this USD difference by 50% to reflect the likelihood that turbine manufacturers do not pass through (and may not even experience) the full extent of exchange rate impacts calculated in this manner (see Appendix C for a discussion of "exchange rate pass-through"). All amounts are expressed in 2010 USD.

The results, shown in Figure 15,[23] suggest that USD weakness added \$136/kW to the average cost of wind turbines in the US through 2008.[24] Since 2008, there has been a \$15/kW reduction as the U.S. dollar has strengthened somewhat. These results can be directly scaled up or down to accommodate different assumptions about exchange rate pass-through – e.g., doubled to reflect 100% pass-through, halved to reflect 25% pass-through (the inherent uncertainty in the exchange rate pass-through fraction is addressed in Appendix C).

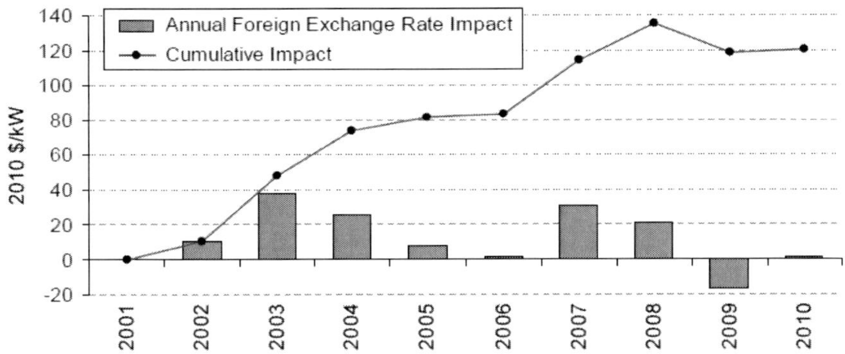

Figure 15. Impact of Currency Movements on Turbine Costs (Annual and Cumulative).

4. Aggregate Impact of Turbine Price Drivers

The individual impacts of each of the seven drivers of wind turbine prices examined in Sections 3.1 through 3.7 are summarized in Table 5. To focus attention on the overall trends, the table presents cumulative impacts over the two periods of major turbine price movements in the past decade – i.e., the doubling in turbine prices from 2002-2008, and the subsequent softening in prices through 2010.

Figure 13, meanwhile, shows yearly impacts, but of all seven drivers combined rather than individually; this aggregate impact (blue line with circle markers) is plotted against the empirical turbine price curve shown earlier in Figure 2 (red line with star markers).

Table 5. Cumulative Impact During Period of Turbine Price Increase (2002-08) and Decrease (2009-10) (2010 $/kW)

	2002-2008		2009-2010	
Endogenous Drivers	+376		-37	
Labor Costs	+91		+12	
Warranty Provisions	+42		-20	
Profit Margins	+59		-78	
Turbine Scaling	+184		+50	
Exogenous Drivers	+219		-53	
Materials Prices	+71		-31	
Steel		*+65*		*-29*
Iron		*+7*		*-2*
Copper		*+9*		*+1*
Aluminum		*+2*		*-1*
Fiberglass		*-11*		*0*
Energy Prices	+12		-7	
Diesel		*+10*		*-4*
Coal		*0*		*0*
Natural Gas		*+2*		*-3*
Currency Movements	+136		-15	
Total Impact	+595		-89	

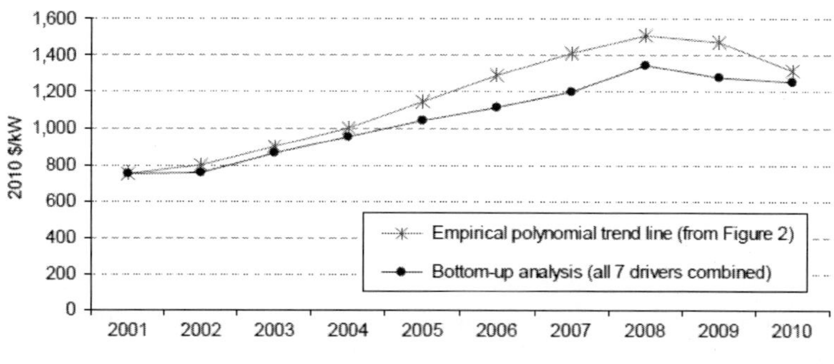

Figure 13. Yearly Impact of All Seven Drivers Combined vs. Empirical Price Curve.

In aggregate, these seven drivers explain nearly $600/kW of the ~$750/kW increase in average turbine prices observed from 2002-2008, and nearly $90/kW of the ~$195/kW decrease in 2009 and 2010 (Table 5, Figure 13). From 2003 through 2010, the bottom-up analysis of these seven drivers explains between 68% and 89% (depending on the year) of the cumulative empirical price movements (Figure 13).[25] Though by no means perfect, this track record improves upon several earlier efforts (Carbon Trust 2008, Greenacre et al. 2010) that have typically failed to quantify more than 60% of observed turbine price increases.

Nonetheless, Figure 13 shows a growing gap between modeled and empirical turbine prices starting in 2005 and increasing through 2007, at which point the gap remains more or less constant until eventually narrowing in 2010. Some portion of this wedge could potentially be explained by our omission of what were likely increasing labor costs and profitability among component suppliers beginning around 2005 (discussed in Section 3.3). The magnitude of the gap – maxing out at $140/kW – is not out of line with changes in turbine manufacturer labor costs and profitability, which rose by a combined $150/kW through 2008 (see Table 5). Alternatively, or in addition, some portion of the gap could reflect higher exchange-rate pass-through than the 50% assumed here, or turbine design and engineering improvements beyond the scaling effects that were analyzed earlier. Finally, some of the discrepancy may simply be due to methodological issues, such as the necessary reliance on Vestas data in this analysis, when the market was supplied by multiple turbine vendors over this period, and especially by GE Wind.

Ignoring these other potential influences not captured here, Table 5 shows that the four partially endogenous drivers (+$376/kW) still account for more of the turbine price increase from 2002- 08 than do the three partially exogenous drivers (+$219/kW). This result suggests that, even absent the considerable exogenous shocks, endogenous price drivers still would have confounded the traditional, simple application of learning curve theory by pushing turbine prices higher even as global wind power installations doubled twice over this period. It is, however, important to note that roughly half of the endogenous impact (+$184/kW) is attributable to turbine up-scaling, for which there is a direct payback in terms of a lower cost of electricity. In other words, from an LCOE perspective, the considerable capital cost impact from turbine up-scaling is not troubling, suggesting that learning effects for wind power should, arguably, be measured through LCOE rather than the more traditionally used turbine prices (or installed project costs). Finally, it is clear from this analysis that there is no single, dominant factor that drove turbine prices higher from 2002-08, or that has yielded lower prices since that time. Turbine up-scaling is, by a significant margin, the largest single driver, although as noted above, the estimated price impact associated with up-scaling can be seen as a reasonable expense given the performance improvements garnered by larger turbines. Currency movements are also found to have played a sizable – though somewhat uncertain – role, as did changes in labor costs and material prices. Changes in manufacturer profit margins, warranty provisions, and energy prices are found to play a less significant, but non-negligible role.

5. LOOKING AHEAD

The analysis described in Section 3, with results summarized and discussed in Section 4, extends through 2010. At the time of writing, however, 2011 is already more than half passed, begging the question of how turbine prices have moved so far in 2011 and how the drivers highlighted in this paper have been impacting those prices. Vestas (2011d) reports that Vestas' average (nominal) price on new orders worldwide during the first half of 2011 was 967 EUR/kW, down 2.5% from 992 EUR/kW in 2010. Bloomberg NEF (2011c) suggests a steeper decline, as the price of turbines within its sample that were contracted in the first half of 2011 for delivery (worldwide) in the coming twelve months fell to 940 EUR/kW, a 7% decline from 2010 levels of around 1,000 EUR/kW. Bloomberg NEF (2011c) goes on to note that the U.S. is a lower cost market than most of Europe, with average pricing pegged at $1,100/kW (i.e., consistent with the 2011 turbine price range presented earlier in Figure 2). Declining labor costs and warranty provisions appear to be enabling some portion of the decline (Vestas 2011d), while ongoing compression of profit margins among turbine manufacturers and component suppliers may be another contributor (Bloomberg NEF 2011c, Hauser 2011, Weiss and Schneeweiss 2011). And, with a presumed continued up-scaling in turbine size in 2011 compared to 2010, the reported price declines have occurred in concert with larger turbines with presumably enhanced performance. At the same time, however, all three exogenous drivers examined in this report – changes in raw materials prices, energy prices, and foreign exchange rates – were pressuring turbine prices higher as of mid-year (Figures 9, 10, and 12 – which show real changes in materials prices, energy prices, and foreign exchange rates, respectively – all extend through June 2011). In addition, as of mid-September 2011, some market participants were noting that turbine prices in the U.S. had begun to move somewhat higher as a result of the rush to have wind turbines under contract before the end of the year in order to qualify for the Section 1603 Treasury cash grant.[26] Thus, it is not yet clear whether the turbine price reductions seen during the first half of 2011 in comparison to 2010 will persist through the second half of the year, or how the full-year 2011 numbers will compare to 2010. The apparent late-year rush induced by the pending expiration of the Section 1603 grant program does, however, highlight another turbine price driver – though of an indirect nature – not yet explicitly discussed in this report: policy risk. The short-term, start-and-stop policy support for wind power that has existed in the U.S. since the production tax credit ("PTC") first expired in mid-1999 created inefficiencies, artificial demand shocks, and sub-optimal investment through at least 2005 (Wiser et al. 2007). In part as a result, in 2005 when the PTC was for the first time extended for two years in advance of expiration, the resulting surge in demand led to major supply bottlenecks, higher labor costs, and rising profit margins – all of which (together with the other endogenous and exogenous cost pressures examined in this report) pushed turbine prices higher. By the time the industry eventually caught up to demand through increased investments in manufacturing and supply chain infrastructure, the global financial crisis of 2008/2009 had wrought significant demand destruction, leaving newly built manufacturing plants operating well below capacity in some cases (Bloomberg NEF 2011c). The Section 1603 program, with its offering of a 30% cash grant in lieu of the PTC, helped to restore demand to a degree, but is now – along with the PTC –also nearing expiration, with no clear guidance as to what, if anything, might replace it. Having ramped up manufacturing capacity in local markets in order to meet demand while also minimizing transport costs and mitigating the risk of adverse exchange rate movements, the industry is now more robust than it was in 2005, which should enable it to focus more on driving the cost of wind energy lower, regardless of the policy environment. Even still, whether the cost of wind energy continues down the long-term downward-sloping cost curve from which it departed in 2002 – but with which it has recently re-engaged – may ultimately depend on what types of policy support are put in place post-2012. Long-term, stable policy support – even if it includes a scheduled ramp down over time to progressively wean the industry off of public support – should enable the industry to capitalize on the investments that it has already made while also planning for the future. A continuation of short-term, stop-and-go policy support, on the other hand, may lead to further rounds of artificial and inefficient demand shocks, with consequent impacts on wind turbine and wind energy pricing.

APPENDIX A – COMMODITY PRICE TIME SERIES DATA

All of the raw materials and energy prices used in this report are priced and traded (in some cases globally) in USD terms. As such, from a U.S.

perspective, these commodities prices are not subject to exchange rate risk, and no exchange rate conversions are required prior to analysis.

Raw Materials Price Data

To enable an assessment of the impact of changes in materials prices on wind turbine prices, this section describes the derivation of the monthly time series of prices for the five materials listed in Table 1 of the main text. Based upon the trends in turbine prices shown earlier in Figure 2, the period of interest for this analysis stretches from the present (i.e., mid-2011) back to the apparent low in turbine prices reached around December 2001.

- *Steel:* We derived a time series of monthly steel prices going back to December 2001 using a combination of price and index data from two independent London-based market analysts and data providers. MEPS (International) Ltd. provides limited historical price (and index) data for a North American Composite Carbon Steel Price, while the CRU Group provides a lengthier history of index (but not price) data for its North American Steel Price Index.[27] As such, our approach was to use the MEPS price for the most recent month available (June 2011), and use the monthly CRU price index to inflate or deflate the June 2011 MEPS price going back through December 2001. During a 4-year period of overlap (July 2007-June 2011), the actual MEPS price data and the derived CRU-index-based price estimates are quite consistent with one another.
- *Fiberglass/Resin/Plastic:* Fiberglass and resin prices were derived from Tables 2.8 and 2.9 of TPI Composites, Inc. (2003), and were assumed to be current as of April 2003. These costs were inflated/deflated over the full period of interest using applicable PPI sub-indices from the Bureau of Labor Statistics.[28] The resulting two time series were combined into a single, material-weighted time series assuming a mix of 60% fiberglass and 40% resin (TPI Composites, Inc. 2003).[29] Plastics are assumed to be closely related to fiberglass and priced the same.
- *Iron/Cast Iron:* The analysis uses historical pig iron prices obtained from steelonthenet.com, as pig iron is a pre-cursor to cast iron (and steel). Not surprisingly, movement in pig iron prices closely reflects

movements in CRU's "Metallics" index (which encompasses scrap iron, pig iron, direct reduced iron, and hot briquetted iron).

- *Copper and Aluminum:* Historical monthly cash settlement prices for both primary aluminum and copper come from the London Metal Exchange ("LME"), one of the world's premier metals markets.
- Each of these five material price time series is converted to \$/kg using standard conversion factors, and normalized to 2010 dollars using the quarterly GDP price deflator published by the United States Department of Commerce's Bureau of Economic Analysis.

Energy Price Data

Monthly time series of fossil fuel prices come from the Energy Information Administration (EIA). We assume that the majority of oil consumption comes in the form of diesel fuel (for transportation), and therefore use the EIA's monthly price estimate for all types of highway diesel fuel.

For natural gas, we used the EIA's average monthly industrial price. Finally, coal prices come from Table 4.1 of the EIA's *Electric Power Monthly* publication, and represent the average cost of coal fuel across all power generation sectors.

For the 40% of energy consumption supplied by fixed-price resources, we simply assume a fixed cost over time (fixed in real terms, escalating at the rate of inflation in nominal terms) of \$100/MWh.

It is perhaps worth noting that none of the three fossil fuels analyzed are traded in a global market (although price movements in various markets around the world are likely highly correlated), and the time series used here are U.S.-specific prices.

Though fossil fuel prices may have been somewhat different in Europe or Asia (where many wind turbines and components were manufactured over the period of interest), it is also the case that foreign turbine manufacturers were steadily increasing their U.S. manufacturing capabilities – and therefore exposure to U.S. energy prices – over this period. In addition, GE Wind – by far the market leader in the U.S. over this period – presumably manufactures most of its turbine components in the U.S., and is therefore paying U.S. energy prices. As such, the use of U.S.-specific energy prices should not impose undue bias.

APPENDIX B – TABLES 6 AND 7 FROM VESTAS LCA OF 1.65 MW V82 TURBINE (VESTAS 2006B)

Table 6. Significant resource consumption of 1kWh electricity from an anshore wind power plant

	Input for 1 kwh electricity delivered from V82-1.65 based onshore wind power plant [kg/kWh]				
	Total	Production	Transport	Operation	Disposal, incl. recovery of metals
Water (fresh) [kg]	3.79E-02	6.76E-02	6.81E-05	1.32E-07	-2.97E-02
Stone [kg]	3.59E-03	3.59E-03	0.00E+00	1.60E-13	3.92E-16
Inert rock [kg]	2.08E-03	1.94E-03	0.00E+00	0.00E+00	1.43E-04
Hard coal [kg]	1.11E-03	2.21E-03	1.48E-06	1.81E-07	-1.11E-03
Iron [kg]	9.94E-04	4.01E-03	2.92E-08	2.39E-10	-3.01E-03
Crude oil [kg]	7.14E-04	5.41E-04	3.03E-04	1.01E-05	-1.40E-04
Natural gas [kg]	5.27E-04	4.87E-04	1.81E-05	1.42E-07	2.09E-05
Limestone [kg]	3.26E-04	3.29E-04	6.39E-08	4.23E-09	-2.36E-06
Lignite [kg]	2.25E-04	2.60E-04	2.66E-08	1.14E-09	-3.58E-05
Sodium chloride (rock salt) [kg]	1.39E-04	1.41E-04	4.18E-08	1.07E-10	-2.40E-06
Quartz sand [kg]	1.23E-04	1.23E-04	4.39E-10	1.25E-09	-5.55E-07
Soil [kg]	3.43E-05	3.43E-05	0.00E+00	0.00E+00	6.53E-08
Kaolin [kg]	1.98E-05	1.98E-05	0.00E+00	0.00E+00	4.13E-11
Gypsum [kg]	1.44E-05	1.44E-05	0.00E+00	0.00E+00	1.55E-09
Dolomite [kg]	1.11E-05	3.42E-05	0.00E+00	0.00E+00	-2.31E-05
Colemanite [kg]	1.10E-05	1.10E-05	0.00E+00	0.00E+00	2.30E-11
Aluminum [kg]	8.26E-06	8.89E-06	2.33E-08	9.54E-11	-6.54E-07

Table 7. Consumption of energy resources for the productionof 1 kWh electricity from an onshore V82-1.65 MW based wind power plant

Non-renewable fuels	kg	MJ/kg	MJ
Crude oil (resource)	7.14E-04	4.57E+01	3.26E-02
Hard coal (resource)	1.29E-03	2.52E+01	3.25E-02
Lignite (resource)	2.29E-04	9.58E+00	2.19E-03
Natural gas (resource)	5.22E-04	5.07E+01	2.65E-02
Uranium (resource)	9.98E-09	6.29E+05	6.27E-03
Renewable fuels	kg	MJ/kg	MJ
Biomass, dry matter, fuel	1.31E-04	2.25E+01	2.94E-03
Primary energy from wind power			1.21E-04
Primary energy from hydro power			5.09E-03
Primary energy from solar energy			9.87E-05
Primary energy from geothermics			3.09E-05
Total (MJ/kWh produced)			1.08E-01
Total (kWh/kWh produced)			3.01E-02
Total (kWh/turbine) in the lifetime			3,392,042

APPENDIX C – EXCHANGE RATE PASS-THROUGH

All else equal, the prolonged period of USD weakness since 2002 has likely inflated the USDdenominated cost of wind turbines imported into the U.S. The magnitude of the impact, however, is less certain. Though not focused on wind turbines, Jabara (2009) notes that the impact of exchange rate movements on the price of finished goods ultimately depends on the exporting firms' price-setting behavior: whether they (1) price to market (i.e., adjust their price to reflect changing conditions in the export market); (2) set their prices in the local currency of the importer (or some third-party benchmark currency, such as the USD); (3) engage in cross-border production (i.e., importing various components – some that might be priced in third-party benchmark currencies, such as the USD – from certain countries, and exporting finished goods to other countries); or (4) some combination of all three. Any of these practices, or combination of practices, could result in less than full pass-through of exchange rate movements.

For example, to remain competitive, an exporting firm that is trying to build or maintain market share might "price to market" by absorbing exchange-rate-related cost increases into its (reduced) profit margin. Although some turbine manufacturers may have selectively employed this strategy for a few years in the early 2000s,[30] ultimately a *sustained* strategy of pricing to market requires finished goods to be sufficiently differentiated such that manufacturers can earn sizeable profit margins – i.e., sizeable enough to be partially sacrificed at times without placing the company at risk (Jabara 2009). Wind turbines do not really fit this mold. Although project developers and financiers will often distinguish between so-called "Tier 1" and "Tier 2" wind turbine manufacturers, there is simply not much variation among wind turbines built by different manufacturers, particularly those that reside within the same tier. As a result, price sensitivity is high, and turbine manufacturer profit margins generally are not (e.g., see Section 3.3 of the main text). This combination – along with rising profit margins from 2005 through 2008 – suggests that although pricing to market (i.e., low pass-through of exchange rate movements) may have been employed early in the study period, any such strategy has likely been largely abandoned in more recent years in favor of rebuilding profitability – and localizing production.[31]

Jabara (2009) goes on to note that "The higher the product differentiation and the lower the price sensitivity, exporters will prefer to export in their own currency." Given the relatively low product differentiation and high price sensitivity surrounding wind turbines, one might therefore expect foreign wind turbines imported into the US to be priced primarily in USD, especially later in the 2001-2010 timeframe – i.e., what Jabara (2009) refers to as "local currency pricing." And this does seem to be the case: indications from one large wind developer in the U.S. are that –with a few exceptions – most turbine supply agreements involving turbines destined for the U.S. market are indeed priced in USD (or are at least mostly priced in USD). On its own, this fact provides little indication of how exchange rate impacts are allocated – i.e., even though a turbine is priced in USD, a turbine manufacturer could still simply raise the USD price to reflect an adverse exchange rate movement. Jabara (2009) relates, however, that exporters who price their goods in the local currency of the importer have, in general, been found to be much less likely (than those who price in their own currency) to pass through the full extent of adverse exchange rate movements.

It is also important to recognize that although Vestas, for example, is based in Denmark, it is sourcing materials and components from many other countries (some of which fall outside the Euro zone), and is also manufacturing finished goods in other countries (again, some of which fall outside of the Euro zone). Furthermore, some of the commodities and other materials that Vestas must purchase in order to manufacture wind turbines may already be priced in USD, due to its status as an international reserve currency. This globalization of the manufacturing process, which Jabara (2009) refers to as "cross-border production," could also limit the extent of exchange rate pass-through, as a given movement in the USD/EUR exchange rate simply may not impact the cost of finished goods on a one-to-one basis.

Although not focused on wind turbines, Jabara (2009) finds – both from her own analysis as well as a review of other studies – very low exchange rate pass-through to the U.S. on average. In fact, exporters seem to be favoring the U.S. in this regard: average pass-through rates to the U.S. are below 0.50 (i.e., less than 50% of exchange rate movements are passed through to U.S. import prices), compared to higher levels in other countries. A high degree of local currency pricing (i.e., more than 90% of U.S. imports are priced in USD) may contribute to this favoritism (Jabara 2009).

Extrapolating Jabara's results to wind turbines is somewhat problematic. On one hand, the fact that most turbine supply agreements are priced in USD (i.e., local currency pricing) would seem to argue for a relatively low pass-through of 0.5 or less (i.e., consistent with the overall situation in the U.S.). On the other hand, the fact that the profit margins of many wind turbine manufacturers generally rose in recent years (at least through 2008) suggests that they are not necessarily sacrificing profit by absorbing adverse exchange rate impacts.[32] Lacking better guidance, our analysis assumes a middle-of-the-road exchange rate pass-through of 0.50; readers interested in examining either higher or lower pass-through can easily scale our results up or down (e.g., for 100% pass-through, simply double our $/kW estimates; for 25% pass-through, cut them in half).

Finally, some mention of currency hedging and how it relates to the foregoing discussion is warranted. In its 2010 Annual Report, Vestas notes that "The business activities of Vestas involve exchange rate risks linked to the purchase and sale of goods and services outside the euro zone. Vestas pursues a policy of hedging exchange rate risks as soon as a commitment in foreign currency is agreed. However, this applies only to net exposure in each individual currency" (Vestas 2011a). It goes on to state that exchange rate risks associated with long-term investments in overseas operations or subsidiaries are not hedged. In other words, Vestas – presumably along with other turbine manufacturers – does hedge short-term exchange rate risks that arise from contractual obligations, but this practice only locks in rates for contracts that have already been executed. Contracts that are still under negotiation, or that will be negotiated some time in the future, will remain subject to price fluctuations stemming from exchange rate movements until such contracts are finalized. Only once a contract has been executed can exchange rate risk be mitigated through hedging. Since the analysis in this report pertains to the impact of exchange rate risk on the price of new contracts signed over the course of a decade, short-term hedging practices are not particularly relevant to this discussion.

REFERENCES

American Wind Energy Association (AWEA). 2011. *U.S. Wind Industry Annual Market Report: Year Ending 2010*. Washington, D.C.: American Wind Energy Association.

Aubrey, Crispin. 2007. "Supply Chain: The race to meet demand." *Wind Directions*, January/February 2007, pp. 27-34.

Berry, D. 2009. "Innovation and the price of wind energy in the US." *Energy Policy*, Volume 37, Issue 11, November 2009, pp. 4493-4499.

Blanco, María Isabel. 2009. "The Economics of Wind Energy." *Renewable and Sustainable Energy Reviews*, vol. 13, pp. 1372-1382.

Bloomberg New Energy Finance ("Bloomberg NEF"). 2011a. *US wind energy services; increased competition in an expanding market.* January 4, 2011

Bloomberg New Energy Finance ("Bloomberg NEF"). 2011b. *North America Wind Market Outlook: Q1 2011*. February 23, 2011.

Bloomberg New Energy Finance ("Bloomberg NEF"). 2011c. *Wind Turbine Price Index, Issue V*. July 27, 2011.

Bloomberg New Energy Finance ("Bloomberg NEF"). 2009. *Q3 Wind Market Outlook*. September 30, 2009.

Bolinger, Mark and Ryan Wiser. 2009. "Wind Power Price Trends in the United States: Struggling to Remain Competitive in the Face of Strong Growth." *Energy Policy*. Vol 37, Issue 3, March 2009, pp. 1061-1071.

BWEA & Garrad Hassan. 2009. *Charting the Right Course: Scenarios for offshore capital costs for the next five years*. BWEA, London.

Carbon Trust. 2008. *Offshore wind power: big challenge, big opportunity*. Carbon Trust, London.

Chupka, Marc W., Gregory Basheda. 2007. *Rising Utility Construction Costs: Sources and Impacts*. Prepared by The Brattle Group for The Edison Foundation.

Dinica, Valentina. 2011. "Renewable electricity production costs – A framework to assist policy-makers' decisions on price support." *Energy Policy*, Volume 39, Issue 7, July 2011, pp. 4153-4167.

D'Souza, N., E. Gbegbaje-Das, P. Shonfield. 2011. *Life Cycle Assessment of Electricity Production from a V112 Turbine Wind Plant*. Prepared for Vestas by PE North West Europe ApS. February 2011.

Efiong, Asari and Andrew Crispin. 2007. *Wind Turbine Manufacturers; Here Comes Pricing Power*. Merrill Lynch Industry Overview, August 10, 2007.

Elsam Engineering A/S. 2004. *Life Cycle Assessment of offshore and onshore sited wind farms*. Document number 200128, translated by Vestas Wind Systems A/S, October 20, 2004.

Ernst & Young. 2009. *Cost of and financial support for offshore wind*. Report prepared for the Department of Energy and Climate Change, April 2009. DECC, London.

European Wind Energy Association (EWEA). 2009. *Wind Energy, the Facts*. Brussels, Belgium: European Wind Energy Association.

Ferioli, F., Schoots, K. & Van der Zwaan, B.C.C. 2009. "Use and limitations of learning curves for energy technology policy: A component-learning hypothesis." *Energy Policy*, Volume 37, Issue 7, July 2009, pp. 2525-2535.

Fingersh, L., M. Hand, A. Laxson. 2006. *Wind Turbine Design Cost and Scaling Model*. NREL/TP-500-40566. Golden, Colo.: National Renewable Energy Laboratory.

Fowler, Andrew. 2008. "Trends in Wind Power Prices, B.O.P., and Turbine Costs." *WINDPOWER 2008*. Houston, Texas, June 2, 2008.

Gamesa. 2011. *Sustainability Report 2010*.

Greenacre, Philip, Robert Gross, Phil Heptonstall. 2010. *Great Expectations: The cost of offshore wind in UK waters – understanding the past and projecting the future*. A report produced by the Technology and Policy Assessment Function of the UK Energy Research Centre.

Hauser, Christine. 2011. "G.E. Profits Are in Line With Analysts' Outlook." *New York Times*, October 21, 2011.

IHS CERA. 2011. "Power Plant Construction Costs: Cost Pressures Returning." Latest report of the *Power Capital Cost Index (PCCI)*, July 6, 2011.

Jabara, Cathy L. 2009. *How Do Exchange Rates Affect Import Prices? Recent Economic Literature and Data Analysis*. United States International Trade Commission, Office of Industries, Publication ID-21, Revised October 2009.

Koomey, J., N.E. Hultman. 2007. "A reactor-level analysis of busbar costs for US nuclear plants, 1970–2005." *Energy Policy*, Volume 35, Issue 11, November 2007, Pages 5630-5642.

Laxson, A., M. Hand, N. Blair. 2006. *High Wind Penetration Impact on U.S. Wind Manufacturing Capacity and Critical Resources*. NREL/TP-500-40482. Golden, Colo.: National Renewable Energy Laboratory.

Martínez, E., F. Sanz, S. Pellegrini, E. Jiménez, J. Blanco. 2009. "Life Cycle Assessment of a Multi-MegaWatt Wind Turbine." *Renewable Energy*, vol. 34, pp. 667-673.

McNerney, J., J.D. Farmer, J.E. Trancik. 2011. "Historical costs of coal-fired electricity and implications for the future." *Energy Policy*, Volume 39, Issue 6, June 2011, Pages 3042- 3054.

Milborrow, David. 2008. "Dissecting Wind Turbine Costs." *WindStats Newsletter*, vol. 21, no. 4, pp. 3-5.

Schleisner, Lotte. 2000. "Life cycle assessment of a wind farm and related externalities." *Renewable Energy*, vol. 20, pp. 279-288.

TPI Composites, Inc. 2003. *Cost Study for Large Wind Turbine Blades: WindPACT Blade System Design Studies*. SAND2003-1428. Albuquerque, New Mexico: Sandia National Laboratories.

U.S. International Trade Commission (USITC). 2009. *Interactive Tariff and Trade Dataweb (Dataweb)*. http://dataweb.usitc.gov

Vestas Wind Systems A/S. 2006a. *Life cycle assessment of offshore and onshore sited wind power plants based on Vestas V90-3.0 MW turbines*. June 21, 2006.

Vestas Wind Systems A/S. 2006b. *Life cycle assessment of electricity delivered from an onshore power plant based on Vestas V82-1.65 MW turbines*. December 29, 2006.

Vestas. 2011a. *Annual report 2010*.

Vestas. 2011b. *Full year 2010 and guidance for 2011: Operating in a market in recovery*. Presentation accompanying the Vestas Annual report 2010.

Vestas. 2011c. *Shareholder information: 1/2011*.

Vestas. 2011d. *Interim financial report, first half year 2011*. Company announcement No. 37/2011, August 17, 2011.

Vestas. 2010a. *Annual report 2009*.

Vestas. 2010b. *New accounting polices for supply-and-installation projects and their effect on 2006-2011*. Company announcement No. 44/2010, November 22, 2010.

Vestas. 2010c. *Implementation of IFRIC 15: Re-state accounts, Translation of guidance for 2010*. November 22, 2010.

Vestas. 2005. *Strategic plan of action 2005-2008*. Stock exchange announcement No. 16/2005, May 26, 2005.

Villadsen, Lars. 2011. Personal communication with Lars Villadsen, Senior Specialist, Investor Relations, Vestas. July 2011.

Waggoner, Tom. 2011. "Riding the Wind: Domesticating the Supply Chain – One Part at a Time." Presentation at *WINDPOWER 2011*, Anaheim, California, May 25, 2011.

Weiss, Richard and Zoe Schneeweiss. 2011. "Siemens Wind-Power Unit Aims to Return to Margin Exceeding 10%." *Bloomberg*, September 29, 2011.

Willow, Christopher, Bruce Valpy. 2011. *Offshore Wind: Forecasts of future costs and benefits*. Report prepared by BVG Associates for RenewableUK.

Wiser, Ryan, and Mark Bolinger. 2011. *2010 Wind Technologies Market Report*. LBNL-4820E. Berkeley, Calif.: Lawrence Berkeley National Laboratory.

Wiser, R., Z. Yang, M. Hand, O. Hohmeyer, D. Infield, P. H. Jensen, V. Nikolaev, M. O'Malley, G. Sinden, A. Zervos. 2011a. "Wind Energy." In IPCC Special Report on Renewable Energy Sources and Climate Change Mitigation [O. Edenhofer, R. Pichs-Madruga, Y. Sokona, K. Seyboth, P.

Matschoss, S. Kadner, T. Zwickel, P. Eickemeier, G. Hansen, S. Schlömer, C. von Stechow (eds)], Cambridge University Press, Cambridge, United Kingdom and New York, NY, USA.

Wiser, R., E. Lantz, M. Bolinger, M. Hand. 2011b (forthcoming). *Recent Developments in the Levelized Cost of Energy from U.S. Wind Power Projects.* Berkeley, Calif.: Lawrence Berkeley National Laboratory.

Wiser, R., M. Bolinger, G. Barbose. 2007. "Using the Federal Production Tax Credit to Build a Durable Market for Wind Power in the United States" *The Electricity Journal.* Vol 20, Issue 9, November 2007, pp. 77-88.

End Notes

[1] Related literature has also sought to explore the historical cost drivers for coal (e.g., McNerney et al. 2011) and nuclear (e.g., Koomey and Hultman 2007).

[2] For example, Vestas categorizes its turbine orders as "supply only" (which includes only delivery and commissioning), "supply-and-installation" (which also includes turbine erection), or "EPC/Turnkey" (which also includes all other BOP and civil construction). Globally, half of all turbine orders placed with Vestas in 2010 were of the "supply-only" variety (Vestas 2011b), while in the U.S., the vast majority of Vestas turbine orders are reportedly for "supply-only" (Villadsen 2011). Data on the market share of EPC contractors from Bloomberg NEF (2011a) suggest that other wind turbine manufacturers are in a similar position – i.e., turbine erection and other BOP work most often falls outside of the turbine supply agreement.

[3] For example, the sample includes just 10 transactions summing to 907 MW announced in 2010 and early 2011 – i.e., just 14% of the 6,280 MW of new turbine orders reported over this period by AWEA (2011). In addition to less transparency surrounding new orders, there have also been fewer orders overall in recent years, partly a function of reduced demand for wind turbines since the financial crisis of 2008/2009. Prior to the crisis, and heading into the peak of the wind turbine market (in terms of demand and pricing), many of the larger U.S. wind project developers entered into multi-year "frame agreements" with turbine manufacturers as a way to secure their anticipated turbine needs for the foreseeable future at a known price. In the wake of the financial crisis, demand for wind power in the U.S. diminished, leaving many developers holding more turbines than they could reasonably deploy, and therefore reducing demand for new turbine orders.

[4] The prices reported by Vestas are derived from order intake (in billion Euros) divided by MW ordered, and represent "averages for all markets, all contract types, all project sizes and all products." As such, unlike the other data points in Figure 2, the Vestas averages are not exclusively U.S.-specific, likely include a higher prevalence of "supply-and-install" and "EPC/Turnkey" contracts, and also presumably include some offshore wind turbines. That said, the fact that the Vestas prices match up quite well with the polynomial trend line from the Berkeley Lab turbine transaction sample suggests that these factors may not unduly influence the averages (e.g., offshore wind still accounts for a relatively small portion of Vestas' turbine sales). Regardless, the Vestas prices are only included here as a benchmarking exercise – i.e., to confirm the trends revealed through our empirical

transaction sample – and are not used in any of the later analysis (although given that much of the later analysis relies on other types of data from Vestas, it is comforting to see the general agreement between the Vestas reported turbine prices and the empirical turbine prices used in the analysis).

[5] It is important to recognize that the wind power industry was not alone in seeing upward pressure on project costs in the years prior to the global financial crisis – other types of power plants experienced similar increases in capital costs. In September 2007, for example, the Edison Foundation published a report showing increases in the installed cost of both natural gas and coal power plants that rival that seen in the wind industry, with cost drivers overlapping those that are highlighted for wind in the present study (Chupka and Basheda 2007). Similarly, the IHS CERA Power Capital Cost Index ("PCCI") of coal, gas, solar, and wind power plants indicates that the average cost of these power plants increased by 90% from 2000 through 2008 (IHS CERA 2011), and have since declined by roughly 10% (though the index has recently begun to creep higher once again). Cost increases for conventional power plants are also covered in Winters (2008).

[6] To clarify, the period of interest to this analysis is 2002-2010. In order to capture changes that took place in 2002, however, we must also rely on data from 2001. As such, 2001 data are reported in some cases throughout the document.

[7] As one example, the price of commodities (raw materials and energy) could be considered somewhat endogenous in that strong demand among wind turbine manufacturers for certain commodities could conceivably increase the price of those commodities. Although recent media attention has focused on just such a demand-driven increase in the price of "rare earth" metals used in a number of "green" technologies, permanent magnet generators (one such technology utilizing rare earths) were not widely deployed in the U.S. wind turbine fleet during the 2002-2010 period of interest to this study. The remaining materials used to manufacture wind turbines are all relatively plentiful, and the wind industry was likely not large enough on its own during the period of interest to impact prices in a meaningful way. In fact, Laxson et al. (2006) find that even under high wind power penetration levels (e.g., with wind power able to meet 20% of total U.S. electricity demand), materials constraints are not likely to be a major issue for wind turbine manufacturers.

[8] In converting to 2010 dollars, we first inflate the reported nominal Euros for all years to 2010 Euros using a Euroarea GDP deflator. Next, we convert the 2010 Euro amounts for all years into 2010 USD using the average 2010 USD/EUR exchange rate. In other words, we convert all years from EUR to USD using a single exchange rate, and as such we do not capture the impact of movements in the exchange rate over time on labor costs. This is by design, since exchange rate impacts are calculated separately later, in Section 3.7. Finally, we use MW delivered rather than MW produced and shipped for two reasons: (1) most importantly, Vestas did not publish data on MW produced and shipped prior to 2005; and (2) as described in the text box on page 10, Vestas has recently shifted its accounting to focus more on deliveries than shipments. That said, we do acknowledge that "MW produced and shipped" is likely a more appropriate denominator than "MW delivered" for measuring per-unit labor costs, given that the production and shipment of turbines is more closely linked (temporally) with the utilization of Vestas labor (i.e., particularly for supply-only contracts, presumably relatively little Vestas labor is utilized once a turbine has shipped, yet the interim between shipment and delivery could be fairly lengthy, due to shipping time and construction delays).

[9] Six Sigma is a quality improvement tool launched by Vestas in 2005 to provide "a common and uniform approach to innovation and process design, with the goal being continuous

improvements that build on the systematic collection of data. In close cooperation with each supplier, Vestas identifies central specifications that are crucial for product reliability and performance. These parameters are monitored in an ongoing process with a view to launching improvement initiatives" (Vestas 2011a). By the end of 2010, Vestas "and the vast majority of its suppliers" had reportedly reached 5 Sigma, up from 4 Sigma at the end of 2008 (Vestas 2011a).

[10] For example, knowing the number and length of blades delivered by a blade manufacturer does not automatically dictate how much nameplate capacity those blades support, since some could be installed on 1.8 MW turbines, others on 2.0 MW turbines, and still others on 3.0 MW turbines (e.g., Vestas offers V90 turbines – i.e., turbines with a 90 meter rotor diameter – rated at all three of these nameplate capacities).

[11] In addition to varying turbine design parameters, the NREL model also allows the user to vary price levels (as proxied by a number of relevant PPI sub-indices) over time. Because we analyze changes in raw materials and energy prices separately in later sections of this paper, here we simply hold the PPI indices fixed at their average 2010 levels for each combination of capacity, hub height, and rotor diameter. By holding the PPI levels constant, we capture just the impact of scaling, which is our intent. That said, we do recognize the historical interdependence between turbine scaling and commodity price impacts – i.e., as turbines scaled up over this period, manufacturing them required increasing amounts of raw materials and energy at the very same time that materials and energy prices were rising. By holding materials and energy prices constant and isolating scaling impacts (as we have done here), and then separately holding turbine mass and energy use constant and isolating materials and energy price change impacts (as we do in later sections), we are not able to capture this interaction between scaling and materials/energy price changes. However, in comparing three different runs of the NREL scaling model – one as described in the text with turbine parameters varying by year and PPI levels fixed at 2010 levels, one with fixed turbine parameters (based on a 1.65 MW turbine) and PPI levels varying by year, and one that varies both turbine parameters and PPI levels over time – we learn two things. First, analyzing commodity price changes and turbine scaling separately yields combined results that likely overstate the true impact, because the 1.65 MW turbine mass used for the commodities price analysis is larger than the average turbines deployed over the majority of the study period (during which commodities prices were rising). Second, the magnitude of the likely overstatement is relatively small – \$25/kW or less – and can therefore be safely ignored.

[12] The columns in Figure 7 are shaded green rather than blue (as in the three earlier figures showing the impact of the previous turbine price drivers) to represent that, in this case, the columns represent the annual change in the driver rather than the driver itself. The red line continues to represent the cumulative change in the driver since 2001.

[13] Several points about the capacity factor and LCOE estimates in Figure 8 merit mention. First, they all reflect an average wind speed of 7.25 m/s at 50 meters (corresponding to a Class IV wind resource) and no subsidies. To the extent that the turbine scaling shown in Figure 6 has been associated with a shift towards the development of lowerwind-speed sites (e.g., Class III rather than Class IV), the increase in capacity factor and decline in LCOE shown in Figure 8 could therefore be over-stated. Second, the increase in capacity factor is due in part to the fact that the average turbine's rotor swept area has increased relative to its nameplate capacity over this period – a change that automatically boosts capacity factor, even absent other scaling or efficiency improvements. Third, the LCOE calculations only capture scaling-related impacts on initial capital costs and energy capture, and do not

account for the potential impacts of scaling on operating costs (i.e., the NREL model simply assumes that O&M costs do not change with scaling). The impact of this omission is unclear: on one hand, replacement parts for larger turbines will likely be more expensive (a positive correlation between scaling and operating costs); but on the other hand, for a given amount of capacity there will be fewer parts to replace, fewer turbines to climb, etc. (a negative correlation between turbine scaling and operating costs).

[14] The handful of raw materials broken out in some LCAs but not listed in Table 1, and that will be excluded from the analysis on the basis of contributing very little mass, do not account for a disproportionate amount of overall turbine costs. Although the value-added from manufacturing these materials into turbine components may result in relatively expensive components (e.g., electronics), the underlying materials themselves are all relatively common and not disproportionately expensive, which suggests that the present analysis of raw materials prices will not be overly biased by excluding them.

[15] In the short term, changes in the price of raw materials may not flow through completely to wind turbine prices, for a number of reasons. Wind turbine manufacturers and their component suppliers may have long-term contracts or price hedges in place that serve to limit the impact of short-term fluctuations in commodity prices. In addition, manufacturers may be willing to absorb (in the case of rising raw materials prices) or reluctant to pass along (in the case of falling materials prices) short-term changes in commodity prices, to gain market share or expand profit margins, respectively. Over the longer term, however, changes in the costs of production will presumably flow through more completely to the price of finished goods. And even over the short term, turbine manufacturers generally prefer to index the price of their turbine supply agreements to commodity price movements. For example, Vestas' 2010 annual report states that Vestas does hedge commodity price risk by entering into long-term contracts, but that "In general, however, Vestas seeks to incorporate commodity price developments into its sales contracts. The final project price typically depends on developments in a number of key parameters, especially commodity prices. Where a customer seeks certainty for the final project price, this is reflected in a premium that compensates Vestas for the risk undertaken" (Vestas 2011a).

[16] Even after these adjustments are made, the resulting numbers will still not be entirely appropriate for our purposes, because the reported energy consumption data for the transport phase also includes turbine erection and interconnection – both of which are assumed to fall outside of the scope of most turbine supply agreements in the U.S. As such, the energy consumption data used in this section likely overstates the impact of energy price changes on wind turbine prices.

[17] Presumably very little nuclear or renewable power is used during the transport and operations stages of the life cycle. Though some may be used during the disposal stage, it is perhaps just as likely – based on the fossil fuel numbers from Table 6 (of Vestas 2006b) – that nuclear and renewable resource consumption during the disposal stage is negative, reflecting a recovery of some of the energy used to manufacture the turbine originally due to materials recycling. On balance, therefore, the most logical course of action is to simply attribute all of the nuclear and renewable power to the production stage. Even if this attribution is not entirely correct, any error only affects 11% of total energy consumption, thereby minimizing its overall impact.

[18] Although the boundaries of the various life-cycle stages can vary by LCA (e.g., some include turbine erection within the transport phase, while others do not), thereby limiting the usefulness of comparisons, several other Vestas LCAs reviewed for this study find similar energy consumption among the first two life-cycle phases (manufacture and transport). For

example, the LCA of the Vestas V90 3.0 MW turbine (Vestas 2006a) found 9.44 GJ/kW, while the more recent LCA of the Vestas V112 3.0 MW (D'Souza et al. 2011) found 10.5 GJ/kW.

[19] Because the wind turbine prices shown in Figure 2 earlier are generally assumed to not include the cost of foundations, concrete was excluded when analyzing raw materials costs in the previous section. LCA data on energy consumption during the transport phase, however, includes the energy required to construct a foundation, as well as the energy embodied in the concrete and steel that comprise the foundation. As such, it is appropriate to include concrete when calculating embodied energy (for this same reason, the mass of steel shown in Table 3 is larger than that shown earlier in Table 1 – the difference is the rebar used in the foundation).

[20] One can arrive at a similar estimate coming from a different direction through a back-of-the-envelope calculation pieced together from various sources. A figure on page 53 of Vestas (2010a) suggests that Vestas factories are responsible for 9% of total primary energy consumption during the life cycle of a Vestas wind turbine, with another 8% attributed to transport and installation and the remaining 83% attributed to raw materials and Vestas' suppliers (unfortunately, this 83% is not broken down between raw materials and suppliers). In combination, Figures 5, 15, and 16 from D'Souza et al. (2011) suggest a similar, though slightly lower, share for Vestas factories – 7.4% of total primary energy consumption. Meanwhile, two different sources (Efiong and Crispin 2007, Aubrey 2007) suggest that Gamesa is roughly 50% vertically integrated – i.e., it makes roughly as many turbine components in-house as it outsources to its suppliers – and one of these sources ranks Vestas (7 out of 10) slightly lower than Gamesa (8 out of 10) on degree of vertical integration (Efiong and Crispin 2007). Thus, if we assume that Vestas factories account for 7.4% of total primary energy consumption (per D'Souza et al. 2011), that Vestas' suppliers account for at least another 7.4% (based on the degree of vertical integration at Vestas being slightly less than Gamesa's 50%, and assuming that the insourcing/outsourcing of turbine components correlates well with related energy consumption), and that transport accounts for an additional 6% (per D'Souza et al. 2011, and close to the 8% reported in Vestas 2010a), then the manufacturing and transport stages together are responsible for at least 21% of total primary energy consumption, or 2.32 GJ/kW based on energy consumption data reported in D'Souza et al. (2011). This outcome is similar in magnitude to the 2.99 GJ/kW arrived at in the text by subtracting embodied energy from LCA data.

[21] Enron Wind, in turn, had bought both Zond (American) and Tacke (German) in 1997. Zond, in turn, had purchased variable-speed technology from Kenetech (formerly known as U.S. Windpower) in 1996 during bankruptcy proceedings.

[22] The numbers in Table 4 capture only the imports related to the Harmonized Tariff Schedules for "wind-powered generating sets" and "towers and lattice masts," and assume that 100% of the former and 95% of the latter are attributable to wind turbines. Though other wind turbine components are also imported into the United States, wind-specific Harmonized Tariff Schedules are not available for those imports, making it difficult to discern from which countries they originated using U.S. Department of Commerce data.

[23] As with Figure 7 earlier, the columns in Figure 15 are shaded green rather than blue to denote that, in this case, they represent the annual change in the driver, rather than the driver itself. The red line continues to represent the cumulative change in the driver since 2001.

[24] Another way to crudely estimate the impact of exchange rate movements (or at least the USD/EUR exchange rate movement) is to start with Vestas' reported revenue from 2001 to 2008 (pre-restatement) expressed in nominal EUR/kW (delivered) terms, and then convert

that revenue to 2010 USD/kW terms in two different ways – one accounting for movements in the USD/EUR exchange rate and the other not. The former converts nominal EUR/kW to nominal USD/kW in each year using year-end USD/EUR exchange rates, and then inflates nominal USD/kW to 2010 USD/kW using a US GDP deflator. The latter inflates nominal EUR/kW to 2010 EUR/kW using a Euro-area GDP deflator, and then converts all years to 2010 USD/kW using only the USD/EUR exchange rate from year-end 2010. The difference between these two approaches – i.e., roughly $280/kW by 2008 – crudely reflects movements in the USD/EUR exchange rate over the period 2002-2008. This $280/kW estimated impact is roughly twice that estimated in the text assuming 50% exchange rate pass-through; an (aggressive) assumption of 100% pass-through would therefore yield a roughly equivalent impact.

[25] If one assumes 80% exchange-rate pass-through (rather than 50%), the explanatory power rises to a range of 77%- 102%. An assumption of more than 80% exchange-rate pass-through on products exported to the U.S. would appear to be aggressive, however, given that commodities prices – most of which are already traded in USD, and are therefore not subject to exchange rate risk – make up roughly 15% of the cost of a wind turbine.

[26] The Section 1603 grant was originally enacted as part of the American Recovery and Reinvestment Act of 2009, and provides a 30% cash grant in lieu of either the Section 48 investment tax credit ("ITC") or the Section 45 production tax credit ("PTC"). In order to qualify for the grant, wind projects must be "under construction" by the end of 2011, and must achieve commercial operations by the end of 2012. Based on "safe harbor" guidelines issued by the U.S. Treasury, one way in which projects can qualify as being "under construction" is to incur more than 5% of eligible project costs by the end of 2011. In many cases, developers have been complying with the 5% safe harbor by contracting for turbines in 2011 with the intent to either take title to or delivery of a sufficient portion of those turbines in order to meet the 5% threshold by the end of the year.

[27] MEPS price data are available at http://www.meps.co.uk/allproducts%20steel%20price.htm; CRU index data are available at http://www.cruonline.crugroup.com/SteelFerroalloys/CRUSteelPrices/CRUspi/tabid/143/Default.aspx.

[28] For fiberglass, we used the PPI Industry Data Series ID PCU3272123272128 ("Glass fiber mat, textile-type, made by establishments producing glass"), and for resin, we used the PPI Industry Data Series ID PCU3255203255204 ("Synthetic resin & rubber adhesives, incl all types of bonding & laminating adhesives").

[29] The reason for combining fiberglass and resin prices into a single time series is that some LCAs provide mass data for both materials, while others (including the Vestas 1.65 MW LCA) simply list a single mass for "fiberglass," which presumably includes both the fiberglass as well as the resin used to bind it. Still others use the word "plastic" when seeming to refer to fiberglass (possibly a translation problem).

[30] For example, after five years of steadily declining profit margins – perhaps in part due to "pricing to market" – Vestas launched its "Will to Win" campaign in late 2005 with the goal of returning the company to healthy profitability through a variety of measures, including price increases (Vestas 2005). Coming at a time when demand for wind turbines began to outpace supply, this price increase was sustainable up until the global financial crisis of 2008/2009 dramatically changed the balance of supply and demand.

[31] For example, Vestas' 2010 annual report states that "In combination with a higher degree of sourcing from countries whose currencies are not linked to the EUR, the [manufacturing] investments in the USA and China ensure an improved currency mix between income and expenses, making Vestas less sensitive to fluctuations in currencies such as the USD

relative to EUR." In other words, given that Vestas cannot afford to hold the line on exchangerate-related cost increases, but will be rendered significantly less competitive if it passes them through, localizing production in large markets to minimize exchange rate risk makes good business sense.

[32] Of course, it is certainly a possibility that, despite their growth, profit margins are still depressed below where they would have been absent adverse exchange rate movements.

INDEX

D

E

N

O

P

Q

R

S

T

U

V

W

Y